Introduction to Analytical Dynamics

Introduction to Analytical Dynamics

N. M. J. WOODHOUSE

CUF Lecturer in Mathematics,
University of Oxford

CLARENDON PRESS • OXFORD
1987

Oxford University Press, Walton Street, Oxford OX2 6DP
Oxford New York Toronto
Delhi Bombay Calcutta Madras Karachi
Petaling Jaya Singapore Hong Kong Tokyo
Nairobi Dar es Salaam Cape Town
Melbourne Auckland
and associated companies in
Beirut Berlin Ibadan Nicosia

Oxford is a trade mark of Oxford University Press

Published in the United States
by Oxford University Press, New York

British Library Cataloguing in Publication Data
Woodhouse, N. M. J.
Introduction to analytical dynamics.
1. Mechanics, Analytic
I. Title
531'.01'515 QA805
ISBN 0-19-853198-2
ISBN 0-19-853197-4 Pbk

Library of Congress Cataloging in Publication Data
Woodhouse, N. M. J. (Nicholas Michael John), 1949–
Introduction to analytical dynamics.
Bibliography: p.
Includes index.
1. Mechanics, Analytic. I. Title.
QA805.W88 1987 531 86–16417
ISBN 0-19-853198-2
ISBN 0-19-853197-4 (pbk.)

Filmset and printed in Northern Ireland at The Universities Press (Belfast) Ltd.

Preface

It may seem odd that Newtonian mechanics should still hold a central place in the university mathematics curriculum. But there are good reasons.

- It is one of the most accurate physical theories ever devised. Three hundred years after the publication of Newton's *Philisophiae naturalis principia mathematica* (1687), we should be surprised not that some of his ideas have been superseded by relativity and quantum theory, but that it is still necessary to exercise great subtlety and scientific ingenuity to detect any error at all in the three laws of motion. Even in the prediction of the orbit of the planet Mercury, for example, which was a crucial failure of the classical theory, the discrepancy* is only one part in 10^7. For the other planets, it is much less.

  Newton's theory is the prime example of what Wigner[1] calls the 'unreasonable effectiveness of mathematics' as a tool for understanding the physical world—an aspect of the truth of mathematics that can easily be lost in a course overburdened with abstraction.

- Quantum theory and relativity have overthrown the classical view of physics, but the mathematical formalism of classical mechanics still plays an essential part. It provides both a framework for interpretation and a first introduction to key ideas and techniques (frames of reference, general coordinate transformations, the connection between observables and symmetries, . . .). It is an essential prerequisite for any advanced course on applications of mathematics in modern theoretical physics.

- It develops geometric intuition and gives invaluable practice in problem solving and mathematical modelling. It is easy to poke fun at the seemingly endless supply of light rods, inextensible strings, and smooth hemispheres. But all undergraduate exercises are necessarily artificial, however cleverly they are dressed. The strength of mechanics is the vast range of its examples—something that their familiarity can make us overlook—and the diversity of different mathematical ideas that they illustrate.

* The radius vector from the Sun to Mercury sweeps out a total angle of 150 000° per century. The prediction of the Newtonian theory is 43″ less than the observed angle.

- The problems of classical mechanics and, in particular, the centuries of work on planetary motion, stimulated the development of much of modern mathematics. It is no coincidence that the great names of mechanics—Newton, Euler, Poisson, Lagrange, Hamilton, . . . —also occur over and over again throughout many branches of pure mathematics. It is essential to study classical mechanics to understand the roots of mathematics.

- The influence of classical mechanics is still present in modern pure mathematical research. The study of Hamilton's equations, for example, led to the development of symplectic geometry, which in turn has found recent applications in the analysis of partial differential equations and in the representation theory of Lie groups.

A glance through the pages that follow will not reveal anything strikingly unfamiliar. The range of topics is central and traditional, partly because I want the book to be short and (OUP willing) cheap, and partly because I intend it to be no more than an introduction. I hope that it will be read in conjunction with the classics and that it will encourage further exploration (in, for example, Arnold's *Mathematical methods of classical mechanics*).[2]

The book is written for second year mathematics undergraduates and assumes familiarity with elementary linear algebra, the chain rule for partial derivatives, and vector mechanics in three dimensions (the last is not absolutely essential). The main intention is, first, to give a confident understanding of the chain of argument that leads from Newton's laws through Lagrange's equations and Hamilton's principle to Hamilton's equations and canonical transformations; and, second, to give practice in problem solving. Most of the exercises and examples are taken from recent Oxford examination papers.

I have concentrated on trying to clarify the points that come up most frequently in tutorials and that I myself found confusing when I first met these ideas. For example: why are you allowed to say that q and $\dot{q}$ are independent? and: why can I not deduce from $\partial L/\partial t = -\partial h/\partial t$ that $h + L$ is independent of t?

It is true, of course, that the most satisfactory way to come to terms with the mathematics of classical mechanics would be to approach the subject from modern differential geometry. But that would mean reducing analytical mechanics to a minority option at the end of the undergraduate course or in the first year of graduate work, which would be a great loss. Instead, I have tried to make use of lessons that I have learnt from differential geometry, but without ever going outside the framework of local, coordinate-based arguments.

I am particularly grateful to Paul Tod and Tom Cooper for many comments on an earlier version of this book; and to Rob Baston, Andy Clark, Mike Dobson, Steve Lloyd, Diana Mountain, Charles Sanderson, and Steve Thorsett for working through the final version.

Oxford N.M.J.W.
1986

Contents

Note: Sections and exercises marked with an asterisk (*) contain harder or less central material that can be omitted. Exercises and examples marked with a dagger (†) have been adapted from Oxford examination questions. Asides and parenthetical remarks are set in smaller type.

1 Frames of reference

1.1 Introduction

The solution of a mechanical problem begins with the kinematic analysis: the analysis of how a system *can* move, as opposed to how it actually *does* move under the influence of a particular set of forces. In this first stage, the essential step is the introduction of coordinates to label the configurations of the system. These might be Cartesian coordinates (for the position of a particle), or angular coordinates (for the orientation of a rigid body), or some complicated combination of distances and angles. The only conditions are that each physically possible configuration should correspond to a particular set of values of the coordinates; and that, conversely, the coordinates should be *independent* in the sense that each set of values of the coordinates should determine a unique configuration.*

The number of coordinates is called the number of *degrees of freedom* of the system.

Example (1.1.1). A particle moving in space has three degrees of freedom. Its position is determined by, for example, three Cartesian coordinates or by three spherical polar coordinates. □

Example (1.1.2). A particle moving on the surface of a sphere has two degrees of freedom, labelled by the two polar angles θ and φ (Fig. 1.1.1). □

Example (1.1.3). Two particles connected by a rigid rod have five degrees of freedom: if the position of one particle is given (three coordinates), then the other can be anywhere on a sphere with its centre at the position of the first particle (two further coordinates). □

Example (1.1.4). A rigid body has six degrees of freedom: three for the position of the centre of mass; two for the direction of some axis fixed in the body; and one for rotations about this axis. □

The second stage (the dynamical part of the problem) is to use

* This informal definition of independence does not bear close scrutiny; but a discussion of the technical issues involved in a full analysis of the concept would not be enlightening at this stage.

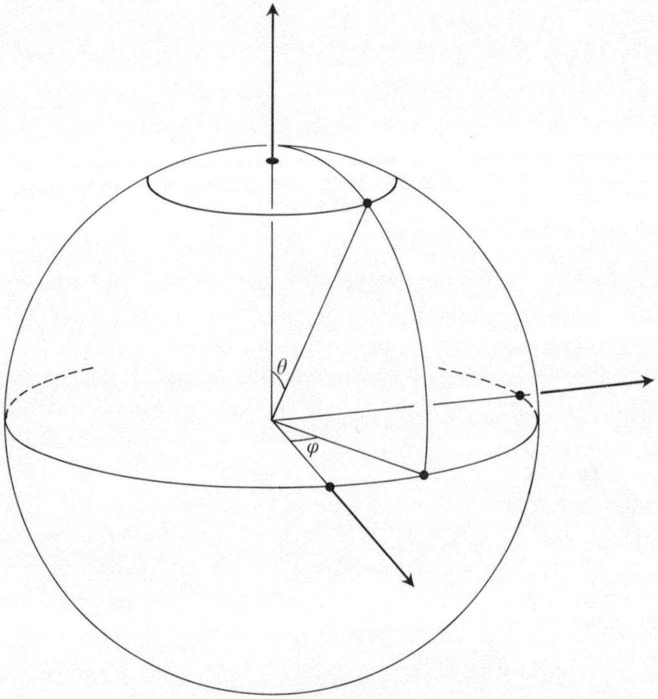

Fig. 1.1.1

Newton's second law to determine the actual motion: to find out how the coordinates evolve as functions of time when the system is subjected to given forces.

In the next few chapters we shall look at a number of techniques for finding and solving dynamical equations in general coordinate systems. These make it possible to simplify the second stage of a variety of mechanical problems, particularly problems involving constraints, by choosing well adapted coordinates in the first stage.

First, however, let us get our bearings by considering a very simple system: a single particle moving in space under the influence of a given force.

The kinematic analysis is easy. We describe the motion of the particle by introducing a frame of reference R, which defines a *standard of rest*.

> **Definition (1.1.1).** A *frame of reference* is an origin together with a set of right-handed Cartesian coordinate axes.

Let $\boldsymbol{r}$ be the particle's position vector from the origin of R. Then the components of $\boldsymbol{r}$ along the axes, r_1, r_2, and r_3, are its Cartesian

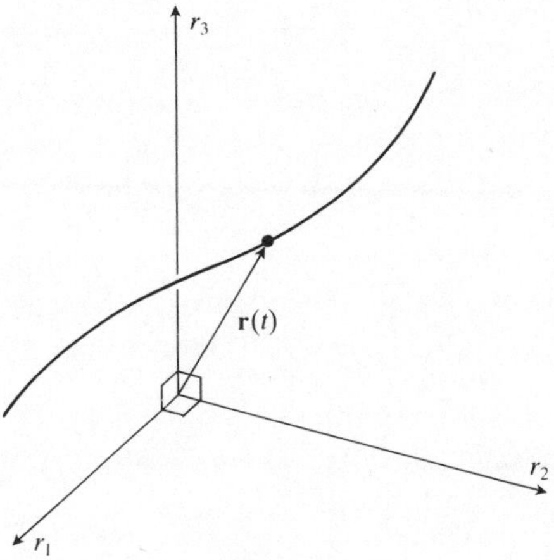

Fig. 1.1.2

coordinates; and a motion of the particle is represented by a curve $\boldsymbol{r}=\boldsymbol{r}(t)$ in space, along which r_1, r_2, and r_3 are functions of time (Fig. 1.1.2).

> **Definition (1.1.2).** The *velocity* and *acceleration* of the particle relative to R are the vectors $\boldsymbol{u}$ and $\boldsymbol{a}$ with components $(\dot{r}_1, \dot{r}_2, \dot{r}_3)$ and $(\ddot{r}_1, \ddot{r}_2, \ddot{r}_3)$, where the dot denotes the derivative with respect to time.

It is important to remember that 'velocity' and 'acceleration' do not make sense unless one adds, either explicitly or by implication, 'with respect to such-and-such a frame of reference'.

Turning to the dynamics, it is an axiom of Newtonian mechanics that there exist special frames of reference, called *inertial frames*, in which Newton's second law holds. If R is such a frame, then

$$m\boldsymbol{a}=\boldsymbol{F} \tag{1.1.1}$$

where m is the mass of the particle and $\boldsymbol{F}=\boldsymbol{F}(\boldsymbol{r}, \boldsymbol{u}, t)$ is the force acting on it, which we shall allow to depend on the position and velocity of the particle, and on the time t. When written out in components, eqn (1.1.1) becomes three simultaneous second-order differential equations,

$$\begin{aligned}\ddot{r}_1 &= F_1(r_1, r_2, r_3, \dot{r}_1, \dot{r}_2, \dot{r}_3, t)\\ \ddot{r}_2 &= F_2(r_1, r_2, r_3, \dot{r}_1, \dot{r}_2, \dot{r}_3, t)\\ \ddot{r}_3 &= F_3(r_1, r_2, r_3, \dot{r}_1, \dot{r}_2, \dot{r}_3, t)\end{aligned} \tag{1.1.2}$$

which determine the three functions $r_i(t)$ $(i = 1, 2, 3)$ in terms of the initial position and velocity.

There is considerable freedom in the choice of the inertial frame. We could pick a different origin or we could choose new directions for the axes. We could also replace R by a second frame $\hat{R}$ moving relative to R without rotation and at constant velocity. Then the particle would have the same acceleration relative to $\hat{R}$ and eqn (1.1.1) would still hold.

For example, if we ignore the effects of the earth's rotation and acceleration, then a set of axes fixed on the earth's surface is an inertial frame. But, again ignoring the effects of rotation and acceleration, Newton's laws are equally valid on the moon (which is moving relative to the earth at about one kilometre per second), or on the sun (about 30 km s^{-1}), or in the Andromeda galaxy (about 270 km s^{-1}). Only an extreme geochauvinist would insist on giving special status to terrestrial frames. As far as mechanical problems are concerned, all non-accelerating, non-rotating frames must be treated equally.

To develop this idea in detail, we need to understand the relationship between coordinates and vector components measured in different frames of reference.

Exercises

(1.1.1) Count the number of degrees of freedom in each of the following systems.

(a) A small bead sliding on a wire.

(b) A lamina moving in its own plane.

(c) A double pendulum confined to a vertical plane. (This consists of a point mass A suspended from a fixed point by a thin rod; and a second point mass B suspended from A by a second thin rod. The rods are hinged at A.)

(d) A double pendulum which is *not* confined to a vertical plane.

1.2 Frames of reference

A frame of reference will be represented as a pair $R = (O, B)$, where O (a moving point) is the origin and $B = (\boldsymbol{e}_1, \boldsymbol{e}_2, \boldsymbol{e}_3)$ is the triad of unit vectors along the coordinate axes (which need not have constant directions). The $\boldsymbol{e}_i$'s must satisfy

$$\boldsymbol{e}_1 \cdot \boldsymbol{e}_1 = \boldsymbol{e}_2 \cdot \boldsymbol{e}_2 = \boldsymbol{e}_3 \cdot \boldsymbol{e}_3 = 1 \tag{1.2.1}$$

at all times since they are unit vectors;

$$\boldsymbol{e}_1 \cdot \boldsymbol{e}_2 = \boldsymbol{e}_2 \cdot \boldsymbol{e}_3 = \boldsymbol{e}_3 \cdot \boldsymbol{e}_1 = 0 \tag{1.2.2}$$

since the axes are orthogonal; and

$$e_1 \cdot (e_2 \wedge e_3) = 1 \tag{1.2.3}$$

since the axes are right-handed.

Equations (1.2.1) and (1.2.2) can be written in a more compact form:

$$e_i \cdot e_j = \delta_{ij} \tag{1.2.4}$$

for $i, j = 1, 2, 3$. Here δ_{ij} is the *Kronecker delta symbol,* defined by

$$\begin{aligned} &\delta_{11} = \delta_{22} = \delta_{33} = 1 \\ &\delta_{12} = \delta_{21} = \delta_{23} = \delta_{32} = \delta_{31} = \delta_{13} = 0. \end{aligned} \tag{1.2.5}$$

Three vectors satisfying eqns (1.2.3) and (1.2.4) (at all times) are said to make up a *right-handed orthonormal triad,* which we shall shorten to 'orthonormal triad', taking 'right-handed' as understood.

Definition (1.2.1). If P is a point, then the vector $\boldsymbol{r}$ from O to P is the *O-position vector* of P; and if $\boldsymbol{x}$ is a vector, then the scalars $x_1 = \boldsymbol{x} \cdot e_1$, $x_2 = \boldsymbol{x} \cdot e_2$, and $x_3 = \boldsymbol{x} \cdot e_3$ are the *B-components* of $\boldsymbol{x}$.

Alternatively, the components can be characterized by

$$\boldsymbol{x} = x_1 e_1 + x_2 e_2 + x_3 e_3. \tag{1.2.6}$$

A vector is not localized at a point: it is simply a quantity with magnitude and direction. If the distance from A to B is the same as the distance from A' to B', and if AB is parallel to $A'B'$, then the vector from A to B is the same as the vector from A' to B'.

Suppose that $B = (e_1, e_2, e_3)$ and $\hat{B} = (\hat{e}_1, \hat{e}_2, \hat{e}_3)$ are two orthonormal triads, which may be rotating relative to each other. Then, for $i, j = 1, 2, 3$,

$$e_i \cdot e_j = \delta_{ij} = \hat{e}_i \cdot \hat{e}_j \tag{1.2.7}$$

(at all times).

Put

$$H_{ij} = e_i \cdot \hat{e}_j \tag{1.2.8}$$

for $i, j = 1, 2, 3$; and

$$H = \begin{pmatrix} H_{11} & H_{12} & H_{13} \\ H_{21} & H_{22} & H_{23} \\ H_{31} & H_{32} & H_{33} \end{pmatrix}. \tag{1.2.9}$$

Definition (1.2.2). H is the transition matrix from $\hat{B}$ to B.

Note the transition matrix from B to $\hat{B}$ is the transposed matrix H^t.

The H_{ij}'s are nine functions of time, labelled by the pairs of indices $i, j = 1, 2, 3$; they determine the relative orientation of the two triads. With i fixed, H_{i1}, H_{i2}, H_{i3} are the $\hat{B}$-components of $\boldsymbol{e}_i$. With j fixed, H_{1j}, H_{2j}, H_{3j} are the B-components of $\hat{\boldsymbol{e}}_j$. Thus for $i = 1, 2, 3$,

$$\begin{aligned} \boldsymbol{e}_i &= H_{i1}\hat{\boldsymbol{e}}_1 + H_{i2}\hat{\boldsymbol{e}}_2 + H_{i3}\hat{\boldsymbol{e}}_3 \\ \hat{\boldsymbol{e}}_i &= H_{1i}\boldsymbol{e}_1 + H_{2i}\boldsymbol{e}_2 + H_{3i}\boldsymbol{e}_3. \end{aligned} \tag{1.2.10}$$

Example (1.2.1). Suppose that

$$\begin{aligned} \hat{\boldsymbol{e}}_1 &= \cos\theta\, \boldsymbol{e}_1 - \sin\theta\, \boldsymbol{e}_2 \\ \hat{\boldsymbol{e}}_2 &= \sin\theta\, \boldsymbol{e}_1 + \cos\theta\, \boldsymbol{e}_2 \\ \hat{\boldsymbol{e}}_3 &= \boldsymbol{e}_3. \end{aligned} \tag{1.2.11}$$

Then

$$H = \begin{pmatrix} \cos\theta & \sin\theta & 0 \\ -\sin\theta & \cos\theta & 0 \\ 0 & 0 & 1 \end{pmatrix} \tag{1.2.12}$$

and

$$\begin{aligned} \boldsymbol{e}_1 &= \cos\theta\, \hat{\boldsymbol{e}}_1 + \sin\theta\, \hat{\boldsymbol{e}}_2 \\ \boldsymbol{e}_2 &= -\sin\theta\, \hat{\boldsymbol{e}}_1 + \cos\theta\, \hat{\boldsymbol{e}}_2 \\ \boldsymbol{e}_3 &= \hat{\boldsymbol{e}}_3. \end{aligned} \tag{1.2.13}$$

The triad $(\boldsymbol{e}_1, \boldsymbol{e}_2, \boldsymbol{e}_3)$ is obtained from $(\hat{\boldsymbol{e}}_1, \hat{\boldsymbol{e}}_2, \hat{\boldsymbol{e}}_3)$ by rotation through θ about an axis parallel to $\hat{\boldsymbol{e}}_3$ (Fig. 1.2.1). □

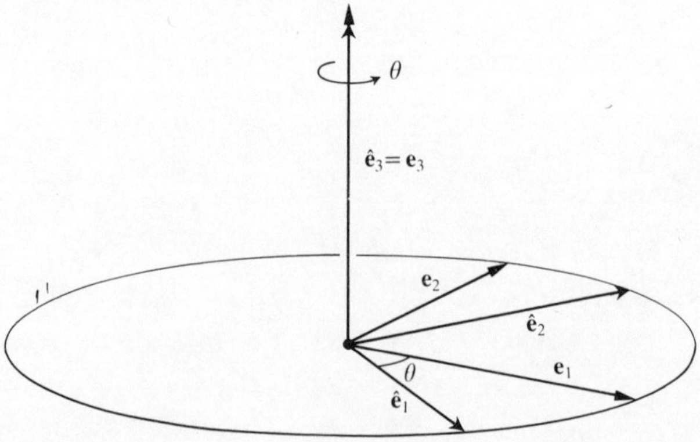

Fig. 1.2.1

Equation (1.2.10) can be written in a more compact form by using the Einstein conventions.

The Einstein conventions. *The summation convention:* whenever an index $i, j, k, \ldots$ is repeated in some term in an expression, a summation over 1, 2, 3 is understood. For example, the definition of the scalar product can be written as

$$\boldsymbol{a} \cdot \boldsymbol{b} = a_i b_i, \tag{1.2.14}$$

where the right-hand side is equal to

$$a_1 b_2 + a_2 b_2 + a_3 b_3 \tag{1.2.15}$$

since the repetition of i implies a sum over $i = 1, 2, 3$. In order to avoid ambiguity, a repeated index must appear exactly twice in each term. An index which is not repeated, and over which there is therefore no summation, is called a *free index.*

The range convention: whenever an index $i, j, k, \ldots$ appears as a free index on both sides of an equation, the equation is understood to hold for all possible values of the index.

The exercises at the end of this section give some practice in the interpretation and manipulation of expressions in which the Einstein conventions are used.

When we adopt the Einstein conventions, eqn (1.2.10) becomes

$$\boldsymbol{e}_i = H_{ij}\hat{\boldsymbol{e}}_j \quad \text{and} \quad \hat{\boldsymbol{e}}_i = H_{ji}\boldsymbol{e}_j. \tag{1.2.16}$$

Now let $\boldsymbol{x}$ be a general vector and put $x_i = \boldsymbol{x} \cdot \boldsymbol{e}_i$ and $\hat{x}_i = \boldsymbol{x} \cdot \hat{\boldsymbol{e}}_i$. The x_i's are the B-components of $\boldsymbol{x}$ and the $\hat{x}_i$'s are the $\hat{B}$-components of $\boldsymbol{x}$. By substituting from eqn (1.2.16), we obtain

$$x_i = \boldsymbol{x} \cdot \boldsymbol{e}_i = H_{ij}\boldsymbol{x} \cdot \hat{\boldsymbol{e}}_j = H_{ij}\hat{x}_j \tag{1.2.17}$$

and

$$\hat{x}_i = \boldsymbol{x} \cdot \hat{\boldsymbol{e}}_i = H_{ji}\boldsymbol{x} \cdot \boldsymbol{e}_j = H_{ji}x_j \tag{1.2.18}$$

or, in matrix notation,

$$\begin{pmatrix} x_1 \\ x_2 \\ x_3 \end{pmatrix} = H \begin{pmatrix} \hat{x}_1 \\ \hat{x}_2 \\ \hat{x}_3 \end{pmatrix} \quad \text{and} \quad \begin{pmatrix} \hat{x}_1 \\ \hat{x}_2 \\ \hat{x}_3 \end{pmatrix} = H^{\mathrm{t}} \begin{pmatrix} x_1 \\ x_2 \\ x_3 \end{pmatrix}. \tag{1.2.19}$$

Since this holds for any $\boldsymbol{x}$, H^{t} must also be the inverse of H. Hence $H^{\mathrm{t}}H = I = HH^{\mathrm{t}}$ where I is the 3×3 identity matrix. In other words, H is an *orthogonal matrix.*

The orthogonality of H also follows directly from eqns (1.2.7) and (1.2.16):

$$\delta_{jk} = \boldsymbol{e}_j \cdot \boldsymbol{e}_k = H_{ji}H_{km}\hat{\boldsymbol{e}}_i \cdot \hat{\boldsymbol{e}}_m = H_{ji}H_{ki}. \tag{1.2.20}$$

Thus $H_{ji}H_{ki} = \delta_{jk}$, which is $HH^t = I$ written with the Einstein conventions. Exercise (1.2.8) contains an outline of a demonstration that the right-handedness of the two triads implies that H is also *proper*; in other words, that $\det(H) = 1$.

We shall now consider how to describe the *rate of rotation* of B relative to $\hat{B}$. By differentiating the orthogonality condition $H_{ji}H_{ki} = \delta_{jk}$,

$$\dot{H}_{ji}H_{ki} + H_{ji}\dot{H}_{ki} = 0, \tag{1.2.21}$$

since the δ_{jk} are constant. Therefore the matrix $\Omega = \dot{H}H^t$ with entries $\Omega_{jk} = \dot{H}_{ji}H_{ki}$ is skew-symmetric ($\dot{H}$ is the matrix whose entries are the time derivatives of the entries in H); that is

$$\Omega = \begin{pmatrix} 0 & \omega_3 & -\omega_2 \\ -\omega_3 & 0 & \omega_1 \\ \omega_2 & -\omega_1 & 0 \end{pmatrix} \tag{1.2.22}$$

for some ω_1, ω_2, and ω_3. The signs have been chosen so that

$$\Omega_{jk} = \varepsilon_{ijk}\omega_i \tag{1.2.23}$$

where ε_{ijk} is the *alternating symbol.*

The alternating symbol. ε_{ijk} is a function of the three indices i, j, and k, which can take values 1, 2, or 3. It vanishes if two or more of the indices are equal. Otherwise, it is equal to 1 if i, j, k is an even permutation of 1, 2, 3 or to -1 if i, j, k is an odd permutation of 1, 2, 3. Thus the non-vanishing ε_{ijk} are

$$\begin{aligned} \varepsilon_{123} = \varepsilon_{231} = \varepsilon_{312} = 1 \\ \varepsilon_{132} = \varepsilon_{213} = \varepsilon_{321} = -1. \end{aligned} \tag{1.2.24}$$

For example, when $j = 2$ and $k = 1$, the only nonzero term in the sum over the repeated index i on the right-hand side of eqn (1.2.23) is $\varepsilon_{321}\omega_3$. But $\varepsilon_{321} = -1$ since 321 is an odd permutation of 123. Hence $\Omega_{21} = -\omega_3$, as in eqn (1.2.22).

The alternating symbol can be used to write the definition of the vector product as

$$\boldsymbol{a} \wedge \boldsymbol{b} = \varepsilon_{ijk}a_ib_j\boldsymbol{e}_k. \tag{1.2.25}$$

For example, the coefficient of $\boldsymbol{e}_2$ is

$$\varepsilon_{132}a_1b_3 + \varepsilon_{312}a_3b_1 = a_3b_1 - a_1b_3. \tag{1.2.26}$$

The exercises at the end of the section give practice in the use of the alternating symbol to manipulate expressions involving vector products.

Definition (1.2.3). The angular velocity of B relative to $\hat{B}$ is the vector $\boldsymbol{\omega} = \omega_i \boldsymbol{e}_i$.

The angular velocity is not localized at a particular point or on a particular axis. It is the 'angular velocity of B relative to $\hat{B}$', *not* 'the angular velocity of B relative to $\hat{B}$ about such-and-such an axis'.

Example (1.2.2). Suppose that H is the matrix in example (1.2.1) (eqn (1.2.12)), where θ is now a function of time. Then

$$\dot{H} = \dot{\theta}\begin{pmatrix} -\sin\theta & \cos\theta & 0 \\ -\cos\theta & -\sin\theta & 0 \\ 0 & 0 & 0 \end{pmatrix} \tag{1.2.27}$$

and so

$$\Omega = \dot{H}H^{\mathrm{t}} = \begin{pmatrix} 0 & \dot{\theta} & 0 \\ -\dot{\theta} & 0 & 0 \\ 0 & 0 & 0 \end{pmatrix}. \tag{1.2.28}$$

Therefore the angular velocity of B relative to $\hat{B}$ is $\boldsymbol{\omega} = \dot{\theta}\boldsymbol{e}_3$. □

If we are given H and $\boldsymbol{\omega}$ at some t, then we can find $\dot{H}$ by rewriting the definition of Ω in the form $\dot{H} = \Omega H$. But H determines the orientation of B relative to $\hat{B}$. Thus the way in which B is rotating relative to $\hat{B}$ is encoded in $\dot{H}$ and hence in $\boldsymbol{\omega}$.

We shall return to this shortly. But first we shall look at another characterization of angular velocity in terms of the *derivative* of a vector with respect to an orthonormal triad.

Definition (1.2.4). The *derivative* of the time-dependent vector $\boldsymbol{x} = x_i \boldsymbol{e}_i$ with respect to $B = (\boldsymbol{e}_1, \boldsymbol{e}_2, \boldsymbol{e}_3)$ is the vector $\mathrm{D}\boldsymbol{x} = \dot{x}_i \boldsymbol{e}_i$.

That is, $\mathrm{D}\boldsymbol{x}$ is obtained from $\boldsymbol{x}$ by taking the components of $\boldsymbol{x}$ in B and differentiating them with respect to time; or alternatively, by differentiating the right-hand side of $\boldsymbol{x} = x_i \boldsymbol{e}_i$, treating the $\boldsymbol{e}_i$'s as constants. For example, if $\boldsymbol{r}$ is the position vector of a particle from the origin of a frame $R = (O, B)$, then $\mathrm{D}\boldsymbol{r}$ and $\mathrm{D}^2\boldsymbol{r}$ are the velocity and acceleration of the particle relative to R.

The derivative of a scalar quantity with respect to time is defined without reference to a particular choice of axes: because one can compare the values of a scalar at two different times without ambiguity, it makes sense to talk about the 'rate of change of a scalar' without adding a qualification.

The derivative of a vector, on the other hand, is a more subtle concept. We know how to compare two vectors $\boldsymbol{a}$ and $\boldsymbol{b}$ at different points of space at a fixed time: they are the same if they have the same magnitude and the same direction. And so there is no difficulty in differentiating vectors with respect to parameters such as arc length along a curve, with the time held fixed. But how do we compare $\boldsymbol{a}$ at time t_1 with $\boldsymbol{b}$ at time t_2? Comparing their lengths is easy, but to compare their directions, we must know how to interpret the statement '$\boldsymbol{a}(t_1)$ and $\boldsymbol{b}(t_2)$ point in the same direction'; and that depends on who is doing the comparing. For example someone on earth might say that $\boldsymbol{a}(t_1)$ and $\boldsymbol{b}(t_2)$ pointed in the same direction if they both happened to point vertically upwards. But then they would not appear to have the same directions to an astronaut standing on the moon because the earth and the moon rotate relative to each other between t_1 and t_2.

Proposition (1.2.1) The Coriolis theorem. The derivatives $\mathrm{D}\boldsymbol{x}$ and $\hat{\mathrm{D}}\boldsymbol{x}$ of $\boldsymbol{x}$ with respect to B and $\hat{B}$ are related by

$$\hat{\mathrm{D}}\boldsymbol{x} = \mathrm{D}\boldsymbol{x} + \boldsymbol{\omega} \wedge \boldsymbol{x} \tag{1.2.29}$$

where $\boldsymbol{\omega}$ is the angular velocity of B relative to $\hat{B}$.

Proof. From eqn (1.2.18), the components of $\boldsymbol{x}$ in $\hat{B}$ and B are related by

$$\hat{x}_i = x_j H_{ji}. \tag{1.2.30}$$

Therefore

$$\begin{aligned} \hat{\mathrm{D}}\boldsymbol{x} &= \dot{\hat{x}}_i \hat{\boldsymbol{e}}_i \\ &= (\dot{x}_j H_{ji} + x_j \dot{H}_{ji})\hat{\boldsymbol{e}}_i \\ &= \dot{x}_j \boldsymbol{e}_j + x_j \dot{H}_{ji} H_{ki} \boldsymbol{e}_k \\ &= \mathrm{D}\boldsymbol{x} + x_j \Omega_{jk} \boldsymbol{e}_k \\ &= \mathrm{D}\boldsymbol{x} + \varepsilon_{ijk}\omega_i x_j \boldsymbol{e}_k \\ &= \mathrm{D}\boldsymbol{x} + \boldsymbol{\omega} \wedge \boldsymbol{x}. \end{aligned} \tag{1.2.31}$$

□

Equation (1.2.29) can be taken as the definition of $\boldsymbol{\omega}$. It determines $\boldsymbol{\omega}$ uniquely since if $\boldsymbol{\omega}'$ is another vector with the property

$$\hat{\mathrm{D}}\boldsymbol{x} = \mathrm{D}\boldsymbol{x} + \boldsymbol{\omega}' \wedge \boldsymbol{x} \tag{1.2.32}$$

for all $\boldsymbol{x}$, then $(\boldsymbol{\omega} - \boldsymbol{\omega}') \wedge \boldsymbol{x} = 0$ for all $\boldsymbol{x}$, and so $\boldsymbol{\omega} = \boldsymbol{\omega}'$. However, some care is needed: without going into the construction of the matrix Ω, it is not obvious that there exists a vector $\boldsymbol{\omega}$ such that eqn (1.2.29) holds for all $\boldsymbol{x}$.

Immediate corollaries of the Coriolis theorem are the following.

Proposition (1.2.2). If the angular velocity of B relative to $\hat{B}$ is $\boldsymbol{\omega}$, then the angular velocity of $\hat{B}$ relative to B is $-\boldsymbol{\omega}$.

Proposition (1.2.3). If B has angular velocity $\boldsymbol{\omega}$ relative to $\hat{B}$ and $\hat{B}$ has angular velocity $\hat{\boldsymbol{\omega}}$ relative to B', then B has angular velocity $\boldsymbol{\omega} + \hat{\boldsymbol{\omega}}$ relative to B'.

Proof. Since $\hat{\mathrm{D}}\boldsymbol{x} = \mathrm{D}\boldsymbol{x} + \boldsymbol{\omega} \wedge \boldsymbol{x}$,

$$\mathrm{D}\boldsymbol{x} = \hat{\mathrm{D}}\boldsymbol{x} - \boldsymbol{\omega} \wedge \boldsymbol{x} \tag{1.2.33}$$

for all $\boldsymbol{x}$; hence the angular velocity of $\hat{B}$ relative to B is $-\boldsymbol{\omega}$.

Similarly, if D' denotes the derivative with respect to B', then, for all $\boldsymbol{x}$,

$$\begin{aligned} \hat{\mathrm{D}}\boldsymbol{x} &= \mathrm{D}\boldsymbol{x} + \boldsymbol{\omega} \wedge \boldsymbol{x} \\ \mathrm{D}'\boldsymbol{x} &= \hat{\mathrm{D}}\boldsymbol{x} + \hat{\boldsymbol{\omega}} \wedge \boldsymbol{x}. \end{aligned} \tag{1.2.34}$$

Therefore

$$\mathrm{D}'\boldsymbol{x} = \mathrm{D}\boldsymbol{x} + (\boldsymbol{\omega} + \hat{\boldsymbol{\omega}}) \wedge \boldsymbol{x} \tag{1.2.35}$$

for all $\boldsymbol{x}$ and hence the angular velocity of B relative to B' is $\boldsymbol{\omega} + \hat{\boldsymbol{\omega}}$. □

Suppose now that B and $\hat{B}$ are along the coordinate axes of two frames of reference $R = (O, B)$ and $\hat{R} = (\hat{O}, \hat{B})$.

Definition (1.2.5). The angular velocity of R relative to $\hat{R}$ is the angular velocity of B relative to $\hat{B}$.

Again, the angular velocity is not localized in space. The definition is of the angular velocity of R relative to $\hat{R}$; and *not* of the 'angular velocity of R relative to $\hat{R}$ about such and such an axis'.

The Coriolis theorem is the key to the interpretation of angular velocity. If $\boldsymbol{x}$ is a vector fixed in R, so that $\mathrm{D}\boldsymbol{x} = 0$, then

$$\hat{\mathrm{D}}\boldsymbol{x} = \boldsymbol{\omega} \wedge \boldsymbol{x}. \tag{1.2.36}$$

Equation (1.2.36) describes the way in which $\boldsymbol{x}$ is varying relative to $\hat{R}$. If we take $\hat{R}$ as the standard of rest and ignore terms of order δt^2, then $\boldsymbol{x}$ appears to change between times t and $t + \delta t$ by the addition of

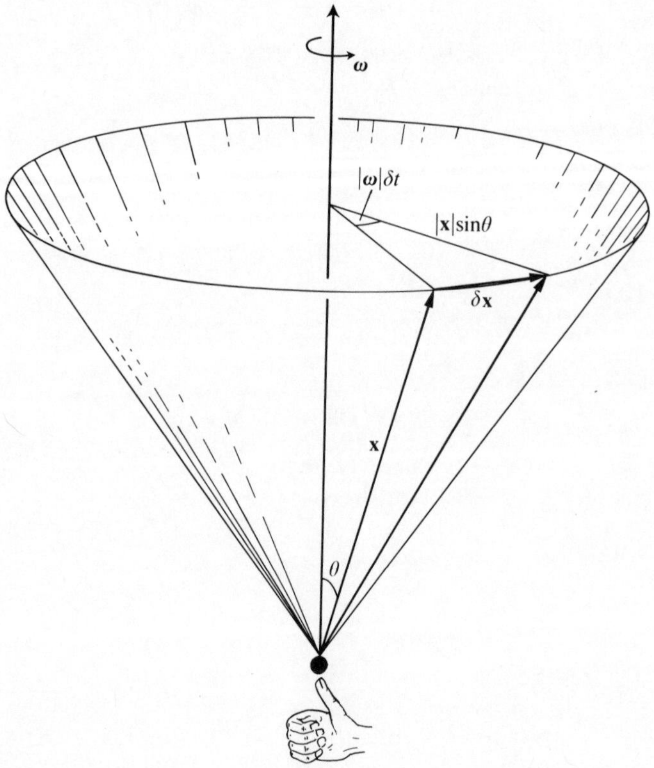

Fig. 1.2.2

$\delta \boldsymbol{x} = \boldsymbol{\omega} \wedge \boldsymbol{x}\delta t$. This is a vector orthogonal to $\boldsymbol{\omega}$ and to $\boldsymbol{x}$ and of modulus $|\boldsymbol{\omega}|\,|\boldsymbol{x}| \sin\theta\,\delta t$, where θ is the angle between $\boldsymbol{\omega}$ and $\boldsymbol{x}$.

Measured in $\hat{R}$, therefore, the change in $\boldsymbol{x}$ between t and $t + \delta t$ is produced by rotating $\boldsymbol{x}$ through an angle $|\boldsymbol{\omega}|\,\delta t$ about an axis parallel to $\boldsymbol{\omega}$, in the right-handed sense (that is: when the thumb of the right hand points along $\boldsymbol{\omega}$, the fingers curl in the direction of the rotation; see Fig. 1.2.2).

Consider now the motion of a particle in the two frames. Let $\boldsymbol{r}$ be the O-position vector of the particle; let $\hat{\boldsymbol{r}}$ be its $\hat{O}$-position vector; and let $\boldsymbol{x}$ be the $\hat{O}$-position vector of O, the vector from $\hat{O}$ to O (see Fig. 1.2.3; remember that $\boldsymbol{r}$, $\hat{\boldsymbol{r}}$, and $\boldsymbol{x}$ all depend on time). Then

$$\hat{\boldsymbol{r}} = \boldsymbol{r} + \boldsymbol{x}. \tag{1.2.37}$$

We need to understand the relationship between $\boldsymbol{a} = \mathrm{D}^2\boldsymbol{r}$, which is the acceleration relative to R, and $\hat{\boldsymbol{a}} = \hat{\mathrm{D}}^2\hat{\boldsymbol{r}}$, which is the acceleration relative

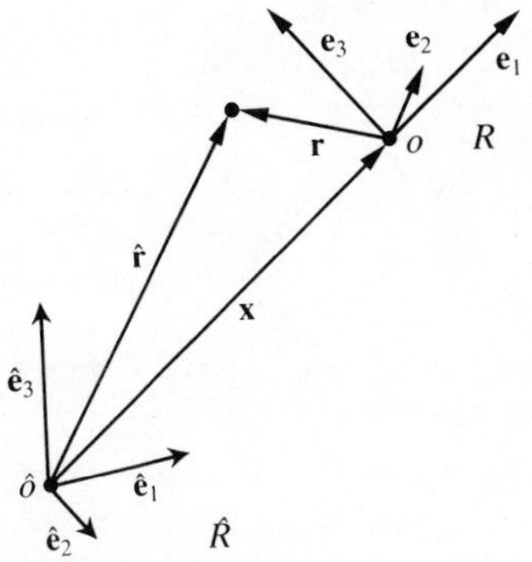

Fig. 1.2.3

to $\hat{R}$. This is found by operating twice with $\hat{\mathrm{D}}$ on eqn (1.2.37):

$$\begin{aligned}\hat{\boldsymbol{a}} &= \hat{\mathrm{D}}^2(\boldsymbol{r}+\boldsymbol{x})\\ &= \hat{\mathrm{D}}(\mathrm{D}\boldsymbol{r}+\boldsymbol{\omega}\wedge\boldsymbol{r})+\boldsymbol{A} \qquad (1.2.38)\\ &= \boldsymbol{a}+(\mathrm{D}\boldsymbol{\omega})\wedge\boldsymbol{r}+2\boldsymbol{\omega}\wedge\mathrm{D}\boldsymbol{r}+\boldsymbol{\omega}\wedge(\boldsymbol{\omega}\wedge\boldsymbol{r})+\boldsymbol{A}\end{aligned}$$

where $\boldsymbol{A} = \hat{\mathrm{D}}^2\boldsymbol{x}$ is the acceleration of O relative to $\hat{R}$ (see exercise (1.2.4)).

We shall leave the full interpretation of eqn (1.2.38) until the next section and consider only the special case in which $\boldsymbol{\omega} = 0$, so that R is not rotating relative to $\hat{R}$ (i.e. the transition matrix from $\hat{B}$ to B is constant), and $\boldsymbol{A} = 0$, so that the point O is moving relative to $\hat{R}$ in a straight line with constant velocity. Then eqn (1.2.38) reduces to $\hat{\boldsymbol{a}} = \boldsymbol{a}$. Under these conditions, therefore, the particle has the same acceleration with respect to the two frames, whatever its motion.

Definition (1.2.6). The two frames R and $\hat{R}$ are *equivalent* if $\boldsymbol{\omega} = 0$ and O is moving relative to $\hat{R}$ in a straight line with constant velocity.

It is easy to see that equivalence of frames is symmetric and transitive: if R is equivalent to $\hat{R}$, then $\hat{R}$ is equivalent to R; and if R is equivalent to $\hat{R}$ and $\hat{R}$ is equivalent to R', then R is equivalent to R'.

The inertial frames are an equivalence class of frames. Newton's second law holds in every inertial frame; and every frame in which it holds (for every choice of $\boldsymbol{F}$ in eqn (1.1.1)) is inertial. The axioms of classical mechanics assert the existence of inertial frames, but they do not single out a particular frame: there is no absolute standard of rest.

There is, however, an absolute standard of acceleration. We can say that a particle has acceleration $\boldsymbol{a}$ if it has acceleration $\boldsymbol{a}$ relative to some, and hence to every, inertial frame. And although it is not possible to test whether or not a frame is 'at rest', it is possible to contrive mechanical experiments that test whether or not a frame is rotating or accelerating. For example, Foucault's pendulum experiment (1851) demonstrated that axes fixed on the earth's surface are not exactly inertial, although the delicacy of the experiment indicated that for most practical purposes they can be treated as such.

Since the axes in two equivalent frames of reference are related by a *constant* orthogonal matrix, the angular velocity of a frame with respect to an equivalent frame is zero. The following is therefore an immediate consequence of proposition (1.2.3).

Proposition (1.2.4). If R has angular velocity $\boldsymbol{\omega}$ relative to $\hat{R}$, then every frame equivalent to R has angular velocity $\boldsymbol{\omega}$ relative to every frame equivalent to $\hat{R}$.

It follows that the angular velocity of a frame relative to an inertial frame is the same, whichever inertial frame is chosen.

Definition (1.2.7). The *angular velocity* of a frame is its angular velocity relative to an inertial frame.

The idea that frames in uniform relative motion are equivalent for describing mechanical phenomena predates Newton's formulation of the laws of classical mechanics. In Galileo's *Dialogue concerning the two chief world systems,* Salviati (who speaks for Galileo) argues, by considering the behaviour of butterflies and fish, that[3] 'in the main cabin below the decks on some large ship' it is impossible to detect the motion of the ship in its effect on physical phenomena so long as 'the motion be uniform and not fluctuating this way and that'. Newton[4] himself stressed the idea. Corollary V to his laws of motion states: 'The motions of bodies included in a given space are the same among themselves, whether that space is at rest or moves uniformly forward in a right line without circular motion'. However, he also asserted the existence of 'absolute space', which remains 'always similar and immovable', and distinguished between absolute motion (that is, motion relative to absolute space) and relative motion.

In the late nineteenth century, the electromagnetic ether was thought to provide an absolute standard of rest, although attempts to detect the effects of motion relative to the ether on electromagnetic processes, such as the experiment of Michelson and Morley (1887), ended in failure. Einstein finally abolished the ether and 'absolute space' in 1905. He asserted the equivalence of inertial frames for describing *all* physical phenomena as a fundamental principle (the *principle of relativity*).

Exercises

(1.2.1) Suppose that the matrix H in eqn (1.2.9) is given by

$$H = \begin{pmatrix} 1/\sqrt{3} & 1/\sqrt{3} & 1/\sqrt{3} \\ -2/\sqrt{6} & 1/\sqrt{6} & 1/\sqrt{6} \\ 0 & -1/\sqrt{2} & 1/\sqrt{2} \end{pmatrix}.$$

Check that $H_{ji}H_{ki} = \delta_{jk} = H_{ij}H_{ik}$. Write down the components of $\boldsymbol{e}_1$, $\boldsymbol{e}_2$, and $\boldsymbol{e}_3$ in $\hat{B}$; and the components of $\hat{\boldsymbol{e}}_1$, $\hat{\boldsymbol{e}}_2$, and $\hat{\boldsymbol{e}}_3$ in B.

(1.2.2) Show that
(i) $\delta_{ij}\delta_{jk} = \delta_{ik}$,
(ii) $\delta_{ij}\delta_{ij} = 3$,
(iii) $\delta_{ii}\delta_{jj} = 9$.

(1.2.3) Let A, B, and C be 3×3 matrices with entries A_{ij}, B_{ij}, and C_{ij} respectively. Show that
(i) $A_{ii} = \mathrm{tr}(A)$,
(ii) $A_{ij}B_{jk}C_{ki} = \mathrm{tr}(ABC)$,
(iii) $A_{ij}\delta_{jp}\delta_{iq}B_{pq} = \mathrm{tr}(AB)$,
(iv) $A_{ij}\delta_{ip}\delta_{jq}B_{pq} = \mathrm{tr}(AB^{\mathrm{t}})$,
(v) $A_{ij}\delta_{pq}\delta_{ij}B_{pq} = \mathrm{tr}(A)\mathrm{tr}(B)$,
where tr denotes trace and B^{t} is the transpose of B.

(1.2.4) Show that
(i) $\mathrm{D}(\boldsymbol{a} \wedge \boldsymbol{b}) = (\mathrm{D}\boldsymbol{a}) \wedge \boldsymbol{b} + \boldsymbol{a} \wedge \mathrm{D}\boldsymbol{b}$,
(ii) $\dfrac{\mathrm{d}}{\mathrm{d}t}(\boldsymbol{a} \,.\, \boldsymbol{b}) = (\mathrm{D}\boldsymbol{a}) \,.\, \boldsymbol{b} + \boldsymbol{a} \,.\, (\mathrm{D}\boldsymbol{b})$.

(1.2.5) Show that
(i) $\varepsilon_{ijk}a_ib_jc_k = \boldsymbol{a} \,.\, (\boldsymbol{b} \wedge \boldsymbol{c})$,
(ii) if $(\boldsymbol{e}_1, \boldsymbol{e}_2, \boldsymbol{e}_3)$ is an orthonormal triad, then

$$\varepsilon_{ijk}\boldsymbol{e}_i \,.\, (\boldsymbol{e}_j \wedge \boldsymbol{e}_k) = 6.$$

(1.2.6) Show that

$$\varepsilon_{ijk}\varepsilon_{klm} = \delta_{il}\delta_{jm} - \delta_{im}\delta_{jl}.$$

Deduce that

$$\boldsymbol{a} \wedge (\boldsymbol{b} \wedge \boldsymbol{c}) = (\boldsymbol{a} \,.\, \boldsymbol{c})\boldsymbol{b} - (\boldsymbol{a} \,.\, \boldsymbol{b})\boldsymbol{c}.$$

(1.2.7) Show that
(i) $\varepsilon_{ijk}\varepsilon_{ijk}=6$
(ii) $\varepsilon_{ijk}\varepsilon_{klm}\varepsilon_{lmn}\varepsilon_{nij}=12$.

(1.2.8) Show that if $\hat{\boldsymbol{e}}_i = H_{ji}\boldsymbol{e}_j$ where $(\boldsymbol{e}_1, \boldsymbol{e}_2, \boldsymbol{e}_3)$ is an orthonormal triad, then

$$\hat{\boldsymbol{e}}_1 \cdot (\hat{\boldsymbol{e}}_2 \wedge \hat{\boldsymbol{e}}_3) = \det(H).$$

Deduce that if $B=(\boldsymbol{e}_1, \boldsymbol{e}_2, \boldsymbol{e}_3)$ and $\hat{B}=(\hat{\boldsymbol{e}}_1, \hat{\boldsymbol{e}}_2, \hat{\boldsymbol{e}}_3)$ are right-handed orthonormal triads, then the transition matrix from $\hat{B}$ to B is a *proper* orthogonal matrix.

(1.2.9) Show that if A is a 3×3 matrix, then

$$\varepsilon_{ijk}A_{i1}A_{j2}A_{k3}=\det(A).$$

Deduce that

$$\varepsilon_{ijk}A_{ip}A_{jq}A_{kr}=\det(A)\varepsilon_{pqr}$$

and hence that if H is a proper orthogonal matrix, then

$$\varepsilon_{ijk}H_{ip}H_{jq}=\varepsilon_{pqr}H_{kr}.$$

(1.2.10)* Use the result of the previous exercise to derive the addition law for angular velocities from the composition rule $H_{ij}=H'_{ik}H''_{kj}$ for transition matrices.

(1.2.11) Show that if H is an orthogonal matrix, then $H^{\mathrm{t}}(H-I)=(I-H)^{\mathrm{t}}$. Deduce that if H is also proper, then $\det(I-H)=0$. Hence show that if B and $\hat{B}$ are two (right-handed) orthonormal triads, then there exists a nonzero vector that has the same components in both triads.

(1.2.12) Show that if ω_i and Ω_{ij} are related by eqn (1.2.23), then $\omega_i\omega_i=\frac{1}{2}\Omega_{ij}\Omega_{ij}$ and $\Omega_{ij}\omega_j=0$.

Show that if there are *constants* x_i such $H_{ij}x_i=x_j$ for all t, then $\boldsymbol{\omega}$ is parallel to $x_i\boldsymbol{e}_i$.

(1.2.13) Show that if H is a proper orthogonal matrix such that $H_{33}=1$, then there is a unique angle $\alpha\in[0, 2\pi)$ such that

$$H=\begin{pmatrix}\cos\alpha & \sin\alpha & 0\\ -\sin\alpha & \cos\alpha & 0\\ 0 & 0 & 1\end{pmatrix}.$$

Show that if $H_{33}=-1$, then there is a unique angle $\alpha\in[0, 2\pi)$ such that

$$H=\begin{pmatrix}-\cos\alpha & -\sin\alpha & 0\\ -\sin\alpha & \cos\alpha & 0\\ 0 & 0 & -1\end{pmatrix}.$$

Sketch a diagram of two orthonormal triads with this transition matrix, showing the angle α.

1.3 Rotating and accelerating frames

In an inertial frame, the dynamical equations governing the motion of a particle are particularly straightforward. Nevertheless, sometimes it is simpler to refer the motion to a non-inertial frame, even though it is then necessary to introduce correction terms to compensate for the frame's acceleration and rotation.

Suppose that $\hat{R} = (\hat{O}, \hat{B})$ is the inertial frame and that $R = (O, B)$ is some other frame. Then,

$$m\hat{\mathbf{a}} = \boldsymbol{F}, \tag{1.3.1}$$

where $\boldsymbol{F}$ is the force acting on the particle, and so

$$m(\mathrm{D}^2\boldsymbol{r} + (\mathrm{D}\boldsymbol{\omega}) \wedge \boldsymbol{r} + 2\boldsymbol{\omega} \wedge \mathrm{D}\boldsymbol{r} + \boldsymbol{\omega} \wedge (\boldsymbol{\omega} \wedge \boldsymbol{r}) + \boldsymbol{A}) = \boldsymbol{F} \tag{1.3.2}$$

by substituting for $\hat{\boldsymbol{a}}$ from eqn (1.2.38).

On transferring terms from one side to the other, this becomes

$$m\mathbf{a} = \boldsymbol{F} - m(\mathrm{D}\boldsymbol{\omega}) \wedge \boldsymbol{r} - 2m\boldsymbol{\omega} \wedge \mathrm{D}\boldsymbol{r} - m\boldsymbol{\omega} \wedge (\boldsymbol{\omega} \wedge \boldsymbol{r}) - m\boldsymbol{A} \tag{1.3.3}$$

which is the same as the equation of motion that we should write down if R were an inertial frame and there were additional forces

$$\begin{aligned} \boldsymbol{F}_1 &= -m(\mathrm{D}\boldsymbol{\omega}) \wedge \boldsymbol{r}, & \boldsymbol{F}_3 &= -m\boldsymbol{\omega} \wedge (\boldsymbol{\omega} \wedge \boldsymbol{r}), \\ \boldsymbol{F}_2 &= -2m\boldsymbol{\omega} \wedge \mathrm{D}\boldsymbol{r}, & \boldsymbol{F}_4 &= -m\boldsymbol{A}, \end{aligned} \tag{1.3.4}$$

acting on the particle. To think of them as real forces, however, is to court disaster: they are nothing more than correction terms that compensate for the acceleration and rotation of R.

The first, $\boldsymbol{F}_1$, arises from the angular acceleration of R relative to $\hat{R}$. The second, $\boldsymbol{F}_2$, is the 'Coriolis force': it is orthogonal to the angular velocity $\boldsymbol{\omega}$ and to the velocity of the particle relative to R. It is responsible, for example, for the circulation of air around an area of low pressure (anticlockwise in the northern hemisphere). The third, $\boldsymbol{F}_3$, is the 'centrifugal force', much exploited in fairgrounds. The fourth, $\boldsymbol{F}_4$, has the form of a uniform gravitational field: it cancels the earth's gravitational field in a free-falling lift or aircraft.

The notation in eqn (1.3.3) is not very convenient for calculation. It is simpler to use a dot for the derivative D with respect to B and to rewrite the equation of motion as

$$m(\ddot{\boldsymbol{r}} + \dot{\boldsymbol{\omega}} \wedge \boldsymbol{r} + 2\boldsymbol{\omega} \wedge \dot{\boldsymbol{r}} + \boldsymbol{\omega} \wedge (\boldsymbol{\omega} \wedge \boldsymbol{r}) + \boldsymbol{A}) = \boldsymbol{F}. \tag{1.3.5}$$

So long as we work entirely in the rotating frame R, and make no further reference to the auxiliary inertial frame $\hat{R}$, then there is no danger of ambiguity.

Example (1.3.1).† A small bead can slide on a smooth wire in the shape of a circle of radius a. The wire is forced to rotate with constant angular velocity $\boldsymbol{\omega}$ about a vertical axis through the centre of the circle. This axis makes an angle α (where $0<\alpha<\pi/2$) with the normal to the plane of the circle. The problem is to find the positions at which the bead can remain at rest relative to the wire.

Choose the rotating frame R so that O is at the centre of the circle and $B=(\boldsymbol{i}, \boldsymbol{j}, \boldsymbol{k})$, where $\boldsymbol{i}=\boldsymbol{e}_1$, $\boldsymbol{j}=\boldsymbol{e}_2$, and $\boldsymbol{k}=\boldsymbol{e}_3$ are fixed relative to the wire: $\boldsymbol{k}$ is normal to the circle, $\boldsymbol{j}$ is horizontal, and $\boldsymbol{\omega}$ lies in the plane spanned by $\boldsymbol{i}$ and $\boldsymbol{k}$ (see Fig. 1.3.1). Then

$$\boldsymbol{\omega}=\omega \sin \alpha\, \boldsymbol{i}+\omega \cos \alpha\, \boldsymbol{k} \tag{1.3.6}$$

where ω is constant; and the position vector of the bead is

$$\boldsymbol{r}=a \cos \theta\, \boldsymbol{i}+a \sin \theta\, \boldsymbol{j}. \tag{1.3.7}$$

The forces on the bead are gravity

$$m\boldsymbol{g}=-mg \sin \alpha\, \boldsymbol{i}-mg \cos \alpha\, \boldsymbol{k} \tag{1.3.8}$$

and the normal reaction $\boldsymbol{N}$ of the wire, which is orthogonal to the wire. Thus the equation of motion is

$$m(\ddot{\boldsymbol{r}}+2\boldsymbol{\omega} \wedge \dot{\boldsymbol{r}}+\boldsymbol{\omega} \wedge(\boldsymbol{\omega} \wedge \boldsymbol{r}))=m\boldsymbol{g}+\boldsymbol{N}, \tag{1.3.9}$$

where the dot is the time derivative with respect to R. By taking the scalar product with

$$\frac{\mathrm{d}\boldsymbol{r}}{\mathrm{d}\theta}=-a \sin \theta\, \boldsymbol{i}+a \cos \theta\, \boldsymbol{j}, \tag{1.3.10}$$

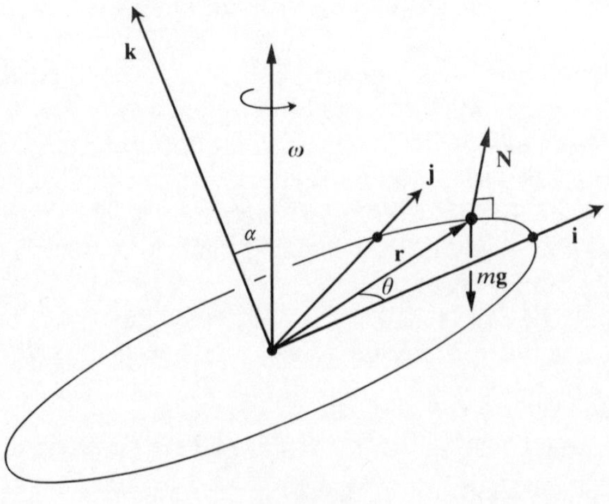

Fig. 1.3.1

which is orthogonal to $\boldsymbol{N}$ and parallel to $\dot{\boldsymbol{r}}$, we obtain

$$\frac{\mathrm{d}\boldsymbol{r}}{\mathrm{d}\theta}\cdot(\ddot{\boldsymbol{r}}+a\omega\cos\theta\sin\alpha\,\boldsymbol{\omega})=ga\sin\alpha\sin\theta. \tag{1.3.11}$$

This reduces to

$$a^2\ddot{\theta}-a^2\omega^2\cos\theta\sin\theta\sin^2\alpha=ga\sin\alpha\sin\theta \tag{1.3.12}$$

on substituting

$$\ddot{\boldsymbol{r}}=-a\ddot{\theta}\sin\theta\,\boldsymbol{i}+a\ddot{\theta}\cos\theta\,\boldsymbol{j}-a\dot{\theta}^2\cos\theta\,\boldsymbol{i}-a\dot{\theta}^2\sin\theta\,\boldsymbol{j}. \tag{1.3.13}$$

The positions of equilibrium (relative to the wire) are given by the solutions of eqn (1.3.12) of the form $\theta=\beta$, where β is constant. The possible values of β are the solutions of

$$\sin\beta(g+a\omega^2\cos\beta\sin\alpha)=0. \tag{1.3.14}$$

If $\omega^2\sin\alpha<g/a$, then there are just two positions ($\beta=0$ and $\beta=\pi$); if $\omega^2\sin\alpha>g/a$, then there are two additional values of β for which

$$\cos\beta=-g/a\omega^2\sin\alpha. \tag{1.3.15}$$

If $\theta=\pi+\varepsilon$, then, to the first order in ε,

$$\ddot{\varepsilon}-\varepsilon\omega^2\sin^2\alpha+\frac{g}{a}\varepsilon\sin\alpha=0, \tag{1.3.16}$$

Hence the lowest position is stable if $\omega^2\sin\alpha<g/a$ and unstable if $\omega^2\sin\alpha>g/a$.

Similarly, if $\theta=\varepsilon$, then, to the first order in ε,

$$\ddot{\varepsilon}-\varepsilon\omega^2\sin^2\alpha-\frac{g}{a}\varepsilon\sin\alpha=0. \tag{1.3.17}$$

Hence the highest position is always unstable.

Example (1.3.2) The rotation of the earth. Suppose that the frame $R=(O,B)$ is fixed relative to the earth, with the origin O on the earth's surface; and that the inertial frame $\hat{R}=(\hat{O},\hat{B})$ has its origin at the centre of the earth (Fig. 1.3.2). Then $\boldsymbol{x}$, the vector from $\hat{O}$ to O, is constant with respect to R since both O and $\hat{O}$ are at rest relative to the earth. Therefore $\ddot{\boldsymbol{x}}=0=\dot{\boldsymbol{x}}$, where the dot denotes the derivative with respect to B, and so

$$\boldsymbol{A}=\boldsymbol{\omega}\wedge(\boldsymbol{\omega}\wedge\boldsymbol{x}). \tag{1.3.18}$$

We shall assume that $\boldsymbol{\omega}$, which is the angular velocity of the earth, is constant with respect to the inertial frame. Then eqn (1.3.5) reduces to

$$m(\ddot{\boldsymbol{r}}+2\boldsymbol{\omega}\wedge\dot{\boldsymbol{r}}+\boldsymbol{\omega}\wedge(\boldsymbol{\omega}\wedge\boldsymbol{r})+\boldsymbol{\omega}\wedge(\boldsymbol{\omega}\wedge\boldsymbol{x}))=m\boldsymbol{g}+\boldsymbol{T}, \tag{1.3.19}$$

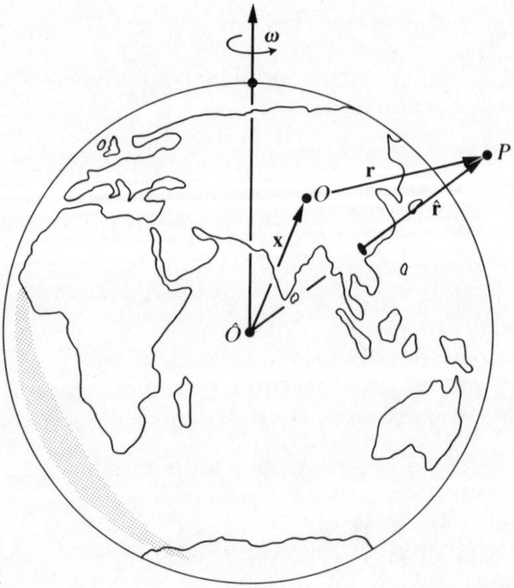

Fig. 1.3.2

where $\boldsymbol{g}$ is the gravitational acceleration and $\boldsymbol{T}$ is the resultant of any other forces acting on the particle.

If the earth were not rotating, then we could measure the gravitational acceleration $\boldsymbol{g}$ by measuring the force $\boldsymbol{T}$ needed to keep the particle at rest relative to the earth; for example, by hanging the particle from a spring and calculating $\boldsymbol{T}$ from the extension. But if we take account of the effects of the earth's rotation, then $\boldsymbol{T}$ is no longer equal to $-m\boldsymbol{g}$: if the particle is at rest relative to R at the point P with O-position vector $\boldsymbol{r}$, then $\dot{\boldsymbol{r}} = \ddot{\boldsymbol{r}} = 0$ and $\boldsymbol{T} = -m\boldsymbol{g}'$, where

$$\boldsymbol{g}' = \boldsymbol{g} - \boldsymbol{\omega} \wedge (\boldsymbol{\omega} \wedge \boldsymbol{x}) - \boldsymbol{\omega} \wedge (\boldsymbol{\omega} \wedge \boldsymbol{r}). \tag{1.3.20}$$

This vector is called the *apparent gravity*.

At the equator the magnitude of $\boldsymbol{g} - \boldsymbol{g}'$ is $3.4 \times 10^{-2}\,\mathrm{m\,s^{-2}}$, which is less than half of one per cent of the true gravitational acceleration.

By substituting eqn (1.3.20) into eqn (1.3.19), we obtain the final form of the equation of motion

$$m(\ddot{\boldsymbol{r}} + 2\boldsymbol{\omega} \wedge \dot{\boldsymbol{r}}) = \boldsymbol{T} + m\boldsymbol{g}'. \tag{1.3.21}$$

The apparent gravity depends on $\boldsymbol{r}$ both through the last term in its definition and through the dependence of $\boldsymbol{g}$ on P. For motion near O, however, we can ignore the variation, and treat $\boldsymbol{g}'$ as constant (with respect to R).

At the earth's surface, the variations in $\boldsymbol{g}$ (and $\boldsymbol{g}'$) over 10 km are of the same order as the difference between real and apparent gravity. This puts a limit on the distance over which it is sensible to take account of the difference between $\boldsymbol{g}$ and $\boldsymbol{g}'$, but to ignore the variations in $\boldsymbol{g}'$.

Example (1.3.3) Foucault's pendulum. Foucault's pendulum gives a concrete application of the equation of motion derived in the previous example. It is named after Léon Foucault's demonstration of the earth's rotation at the Paris Exhibition in 1851 and it consists simply of a bob of mass m suspended from a fixed point by a long wire of length a. It is so constructed that, when set in motion, it will swing for long enough for the Coriolis term $2m\boldsymbol{\omega} \wedge \dot{\boldsymbol{r}}$ in eqn (1.3.21) to have an appreciable cumulative effect (in practice, additional energy is usually provided to keep the pendulum swinging).

Let the origin O of R be the point of suspension; and, to simplify the notation, denote the unit vectors along the axes of R as $\boldsymbol{i}$, $\boldsymbol{j}$, and $\boldsymbol{k}$. We shall suppose that these have been chosen so that $\boldsymbol{k}$ is along the (apparent) vertical and $\boldsymbol{i}$ is a horizontal vector pointing due north. Then,

$$\begin{aligned} \boldsymbol{g}' &= -g'\boldsymbol{k} \\ \boldsymbol{\omega} &= \Omega(\cos\lambda\,\boldsymbol{i} + \sin\lambda\,\boldsymbol{k}) \qquad (1.3.22) \\ \boldsymbol{r} &= r\cos\theta\,\boldsymbol{i} + r\sin\theta\,\boldsymbol{j} + z\boldsymbol{k}, \end{aligned}$$

where Ω is the angular speed of the earth ($2\pi/24$ radians per hour), λ is the apparent latitude (which is defined to be the angle between $\boldsymbol{g}'$ and the equatorial plane), and r, θ, and z are the cylindrical polar coordinates of the bob (see Figs. 1.3.3 and 1.3.4; note that $z < 0$ in the configuration shown).

The equation of motion is

$$m(\ddot{\boldsymbol{r}} + 2\boldsymbol{\omega} \wedge \dot{\boldsymbol{r}}) = -mg'\boldsymbol{k} + \boldsymbol{T}, \qquad (1.3.23)$$

where $\boldsymbol{T}$ is the tension in the wire and $\boldsymbol{r}$ is subject to the constraint $\boldsymbol{r}\,.\,\boldsymbol{r} = a^2$, which fixes the length of the wire. Note that the constraint implies that $\boldsymbol{r}\,.\,\dot{\boldsymbol{r}} = 0$, so that $\dot{\boldsymbol{r}}$ is orthogonal to $\boldsymbol{r}$ and hence also to $\boldsymbol{T}$ (which is along the wire). By taking the scalar product of eqn (1.2.23) with $\dot{\boldsymbol{r}}$, and by cancelling m, we obtain

$$\dot{\boldsymbol{r}}\,.\,\ddot{\boldsymbol{r}} = -g'\boldsymbol{k}\,.\,\dot{\boldsymbol{r}} = -g'\dot{z} \qquad (1.3.24)$$

which, on integration with respect to time, yields the *energy conservation equation* (relative to R)

$$\tfrac{1}{2}\dot{\boldsymbol{r}}\,.\,\dot{\boldsymbol{r}} + g'z = E \qquad (1.3.25)$$

in which E is a constant.

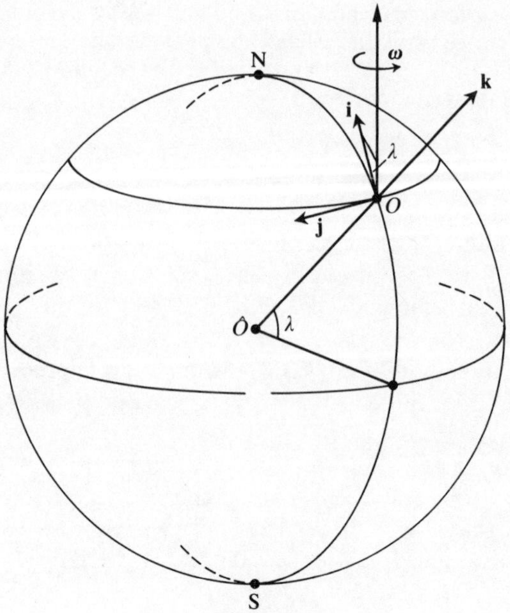

Fig. 1.3.3

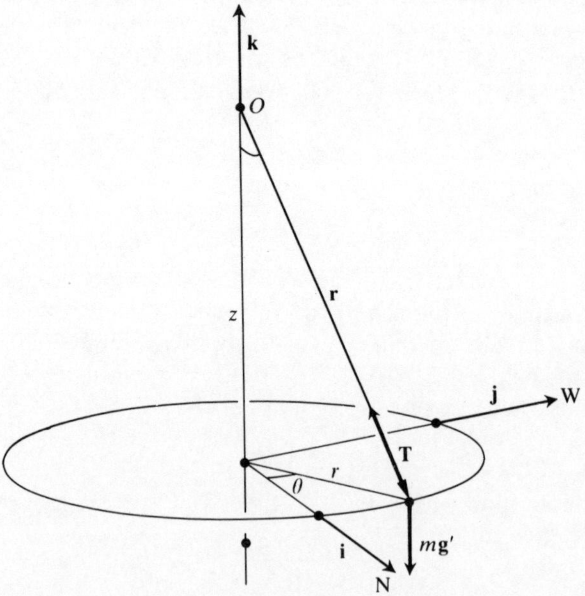

Fig. 1.3.4

Similarly, by taking the vector product with $\boldsymbol{r}$ and then the scalar product with $\boldsymbol{k}$, we obtain the *angular momentum equation*

$$\dot{h} - 2(\boldsymbol{k}\,.\,\dot{\boldsymbol{r}})(\boldsymbol{r}\,.\,\boldsymbol{\omega}) = 0 \tag{1.3.26}$$

in which $h = \boldsymbol{k}\,.\,(\boldsymbol{r} \wedge \dot{\boldsymbol{r}})$. Now

$$\dot{\boldsymbol{r}} = \dot{r}(\cos\theta\,\boldsymbol{i} + \sin\theta\,\boldsymbol{j}) + \dot{\theta}(-r\sin\theta\,\boldsymbol{i} + r\cos\theta\,\boldsymbol{j}) + \dot{z}\boldsymbol{k}, \tag{1.3.27}$$

and by substituting this into eqns (1.3.26) and (1.3.25), we obtain

$$\dot{h} = 2\dot{z}\Omega(r\cos\theta\cos\lambda + z\sin\lambda) \tag{1.3.28}$$

$$\tfrac{1}{2}\left(\dot{r}^2 + \dot{z}^2 + \frac{h^2}{r^2}\right) + g'z = E \tag{1.3.29}$$

together with $h = r^2\dot{\theta}$.

The two equations (1.3.28) and (1.3.29), together with the constraint $z^2 + r^2 = a^2$, determine r, θ, and z in terms of the initial position and velocity. The second equation (1.3.29) does not involve the latitude explicitly.

When $\lambda = \pi/2$, so that the pendulum is over the North Pole, the first equation (1.3.28) reduces to

$$\dot{h} = 2z\dot{z}\Omega. \tag{1.3.30}$$

In this case, we know what happens without solving the equations: O is also at rest relative to the inertial frame and the earth simply turns under the swinging pendulum: the rotation has no other effect on the motion. So, for example, if the bob is pulled to one side and released, then the pendulum swings in a vertical plane; but this plane rotates relative to the earth through a complete revolution once every twenty-four hours, in a clockwise direction.

For a general value of λ, eqn (1.3.28) is

$$\dot{h} = 2z\dot{z}\Omega' + 2\dot{z}r\Omega\cos\theta\cos\lambda \tag{1.3.31}$$

where $\Omega' = \Omega\sin\lambda$. Apart from the replacement of Ω by Ω', this differs from eqn (1.3.30) only by the term $2\dot{z}r\Omega\cos\theta\cos\lambda$. For small oscillations, for which $r/a \ll 1$, this is less significant than the first term on the right-hand side since $|r/z|$ is then much less that 1. If we neglect this term, then we obtain an equation identical to (1.3.30), but with Ω replaced by Ω'. At a general latitude, therefore, the motion is the same as at the North Pole, except that the rate of rotation is $\Omega\sin\lambda$, rather than Ω. For oscillations in a vertical plane at 52°N, for example, the plane of oscillation turns through a complete revolution once every 30·5 hours.

Although we have not considered the motion in any detail, this general discussion has been sufficient to derive the result on which Foucault based his demonstration. A more thorough analysis, such as is given in Synge and Griffith's *Principles of mechanics*,[5] reveals, however, a serious difficulty: there is another effect which causes a rotation of the plane of oscillation and which can completely mask the effect of the Coriolis term.

Without solving the full equations, we can see the origin of the second effect by using phase-plane techniques. We shall come back to this in example (2.2.4).

Exercises

(1.3.1)† A small bead is threaded on a smooth wire in the shape of a curve given parametrically by $\boldsymbol{r}=\boldsymbol{r}(q)$. The wire rotates with constant angular velocity $\boldsymbol{\omega}$ about a vertical axis. Show that

$$\frac{\mathrm{d}}{\mathrm{d}t}\left[\tfrac{1}{2}\dot{q}^2\frac{\mathrm{d}\boldsymbol{r}}{\mathrm{d}q}\cdot\frac{\mathrm{d}\boldsymbol{r}}{\mathrm{d}q}\right]=\dot{q}\frac{\mathrm{d}\boldsymbol{r}}{\mathrm{d}q}\cdot\ddot{\boldsymbol{r}},$$

where the dot is the time derivative with respect to a frame rotating with the wire, and $\boldsymbol{r}$ is measured from an origin on the axis.

Deduce that

$$\frac{\mathrm{d}\boldsymbol{r}}{\mathrm{d}q}\cdot\frac{\mathrm{d}\boldsymbol{r}}{\mathrm{d}q}\dot{q}^2-(\boldsymbol{\omega}\wedge\boldsymbol{r})\cdot(\boldsymbol{\omega}\wedge\boldsymbol{r})-2\boldsymbol{g}\cdot\boldsymbol{r}=\text{constant}.$$

Does this result still hold when the axis of rotation is not vertical?

(1.3.2) A plastic ball is held at the bottom of a bucket of water and then released. As it is released, the bucket is dropped over the edge of a cliff. What happens?

(1.3.3) In example (1.3.1), investigate the stability of the two additional equilibrium points in the case $\omega^2 \sin\alpha > g/a$.

(1.3.4)† A pendulum consists of a light rod and a heavy bob. Initially it is at rest in vertical stable equilibrium. The upper end is then made to move down a straight line of slope α (with the horizontal) with constant acceleration f. Show that in the subsequent motion, the pendulum just becomes horizontal if $g=f(\cos\alpha+\sin\alpha)$.

1.4 The kinematics of rigid bodies

A collection of particles make up a *rigid body* if there exists a frame R relative to which all the particles are at rest at all times. Such a frame is called a *rest frame* of the body. In general, rest frames are *not* inertial.

Equivalently, a rigid body is a collection of particles separated by fixed distances. By taking a limit as the number of particles goes to infinity, we can also think of a rigid body as a continuous distribution of matter, with the separation between different elements remaining constant as the body moves.

It would be an unrewarding mathematical exercise to spell out the precise nature of the limiting process. In any case, both the 'finite collection of particles' and the 'continuous distribution of matter' models are of limited validity.

Provided that the particles do not all lie along a line, the rest frame is determined by the particles up to a rotation of the axes by a constant orthogonal transformation and a translation of the origin by a constant vector (that is, constant in R). In particular, all the rest frames of a general rigid body are equivalent.

Let $R=(O, B)$ be a rest frame and let $\hat{R}=(\hat{O}, \hat{B})$ be some other frame.

Definition (1.4.1). The *angular velocity* $\boldsymbol{\omega}$ of the body *relative to* $\hat{R}$ (or simply the *angular velocity* if $\hat{R}$ is inertial) is the angular velocity of R relative to $\hat{R}$.

Let P be a point fixed in the body with O-position vector $\boldsymbol{r}$. Then its $\hat{O}$-position vector is $\hat{\boldsymbol{r}}=\boldsymbol{x}+\boldsymbol{r}$, where $\boldsymbol{x}$ is the vector from $\hat{O}$ to O.

Let $\boldsymbol{v}_P=\hat{\mathrm{D}}\hat{\boldsymbol{r}}$ and $\boldsymbol{v}_O=\hat{\mathrm{D}}\boldsymbol{x}$ be the velocities of P and O relative to $\hat{R}$. Since O and P are at rest relative to R, $\mathrm{D}\boldsymbol{r}=0$. Therefore

$$\boldsymbol{v}_P=\hat{\mathrm{D}}\boldsymbol{x}+\hat{\mathrm{D}}\boldsymbol{r}=\boldsymbol{v}_O+\boldsymbol{\omega}\wedge\boldsymbol{r}. \tag{1.4.1}$$

But O can be any point of the body. Hence we have the following proposition.

Proposition (1.4.1). Let O and P be two particles in a rigid body and let $\boldsymbol{r}$ be the vector from O to P. Then the velocities $\boldsymbol{v}_O$ and $\boldsymbol{v}_P$ of O and P relative to a frame $\hat{R}$ are related by

$$\boldsymbol{v}_P=\boldsymbol{v}_O+\boldsymbol{\omega}\wedge\boldsymbol{r} \tag{1.4.2}$$

where $\boldsymbol{\omega}$ is the angular velocity of the body relative to $\hat{R}$.

At any instant, therefore, the motion of the body relative to $\hat{R}$ is completely specified by two vectors: the angular velocity $\boldsymbol{\omega}$ and the velocity of any one particle.

We shall use eqn (1.4.2) to define $\boldsymbol{v}_P$ at points outside the body.

Definition (1.4.2). A line L is an *instantaneous axis of rotation* at time t if $\boldsymbol{v}_P(t)=0$ for every $P\in L$.

If one can find an instantaneous axis L (either in the body or in its imaginary extension), then it is easy to picture the motion: the points on L are at rest relative to R at time t and the body is 'rotating about L'.

In general, however, every point of the body is moving relative to R and there is no instantaneous axis, as the following argument shows.

From eqn (1.4.2),

$$\boldsymbol{v}_P \cdot \boldsymbol{\omega} = \boldsymbol{v}_O \cdot \boldsymbol{\omega}. \tag{1.4.3}$$

Therefore $\boldsymbol{v}_P \cdot \boldsymbol{\omega}$ takes the same value at every point. If the body is set in motion with $\boldsymbol{v}_O$ not orthogonal to $\boldsymbol{\omega}$, then $\boldsymbol{v}_P$ cannot vanish for any P.

The converse is also true: if $\boldsymbol{v}_O \cdot \boldsymbol{\omega} = 0$ and $\boldsymbol{\omega} \neq 0$, then the line L given in terms of a parameter λ by

$$\boldsymbol{r} = \frac{1}{\boldsymbol{\omega} \cdot \boldsymbol{\omega}} \boldsymbol{\omega} \wedge \boldsymbol{v}_O + \lambda \boldsymbol{\omega} \tag{1.4.4}$$

is an instantaneous axis.

The line L given by eqn (1.4.4) is still significant even when $\boldsymbol{v}_O \cdot \boldsymbol{\omega} \neq 0$. It is characterized by the property: for any P on L, $\boldsymbol{v}_P$ is proportional to $\boldsymbol{\omega}$. In general, therefore, one can picture the body as having a 'corkscrew' motion, moving along L as it rotates about it. If $\boldsymbol{v}_O \cdot \boldsymbol{\omega} > 0$, then $\boldsymbol{v}_P$ is parallel to $\boldsymbol{\omega}$ on L and the corkscrew is right-handed: if the thumb of the right hand points in the direction of $\boldsymbol{v}_P$, then the fingers curl in the sense of the rotation. Similarly, if $\boldsymbol{v}_P \cdot \boldsymbol{\omega} < 0$, then $\boldsymbol{v}_P$ is anti-parallel to $\boldsymbol{\omega}$ and the corkscrew is left-handed (Fig. 1.4.1).

Definition (1.4.3). A motion of a rigid body is right-handed (left-handed) relative to $\hat{R}$ if $\boldsymbol{v} \cdot \boldsymbol{\omega} > 0$ ($\boldsymbol{v} \cdot \boldsymbol{\omega} < 0$), where $\boldsymbol{v}$ is the velocity of any point of the body.

Example (1.4.1) Rolling conditions. If two rigid bodies are in contact at a point P and if there is no slipping, then the particles of the two bodies at P must have the same velocities.

For example, consider a sphere of radius a rolling without slipping on a rough plane. Let $\boldsymbol{v}$ be the velocity of the centre and let $\boldsymbol{\omega}$ be the angular velocity of the sphere (both relative to a frame fixed on the plane). Then the velocity of the particle of the sphere in contact with the plane is $\boldsymbol{v} + \boldsymbol{\omega} \wedge (-a\boldsymbol{k})$, where $\boldsymbol{k}$ is the unit normal to the plane. The corresponding particle of the plane is at rest. Hence we have the rolling condition

$$\boldsymbol{v} - a\boldsymbol{\omega} \wedge \boldsymbol{k} = 0. \tag{1.4.5}$$

There is a potential confusion here: eqn (1.4.5) is the condition that the *particle* of the sphere in contact with the plane should be instan-

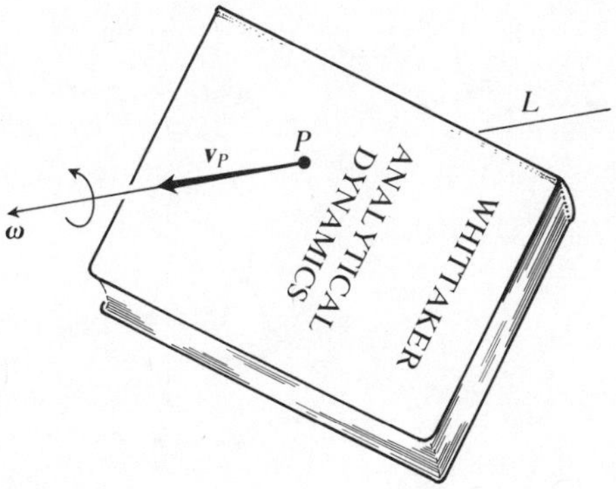

(a)

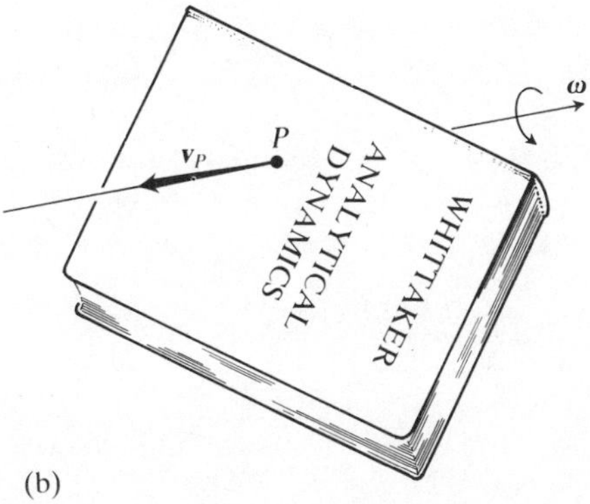

(b)

Fig. 1.4.1 (a) Right-handed motion, (b) left-handed motion.

taneously at rest. However, as the sphere rolls, different particles come into contact and the *point* of contact moves both relative to the sphere and relative to the plane (Fig. 1.4.2).

Example (1.4.2)† A rolling problem. A rough hollow sphere S of radius $3a$ is free to rotate about its centre O. Inside S are five rough solid spheres S_0, S_1, S_2, S_3, S_4, each of radius a (Fig. 1.4.3). Initially the centre of S_0 is at O and the centres of S_1, S_2, S_3, and S_4 are at the vertices of a regular tetrahedron; S_1, S_2, S_3, and S_4 are in contact with the inner surface of S and with the outer surface of S_0. No slipping occurs.

The problem is to show that if S is forced to rotate about O with variable angular velocity, then at all times during the subsequent motion

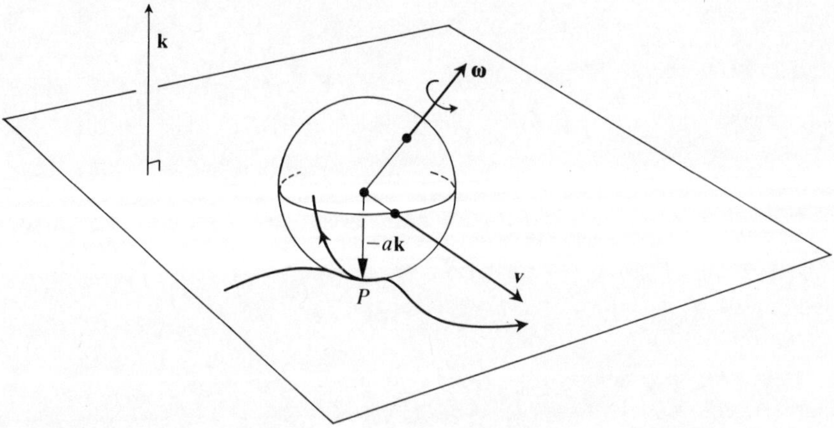

Fig. 1.4.2 Rolling: the curves are the loci of the point of contact on the plane and on the sphere.

the centres of S_1, S_2, S_3, and S_4 form the vertices of a regular tetrahedron.

The first step is to establish the following: suppose that n moving points have position vectors $\boldsymbol{r}_1(t), \boldsymbol{r}_2(t), \ldots, \boldsymbol{r}_n(t)$ from the origin of some frame $\hat{R}$. If there exists a *single* time-dependent vector $\boldsymbol{\theta}(t)$ such that

$$\dot{\boldsymbol{r}}_i = \boldsymbol{\theta} \wedge \boldsymbol{r}_i, \quad i = 1, 2, \ldots, n \tag{1.4.6}$$

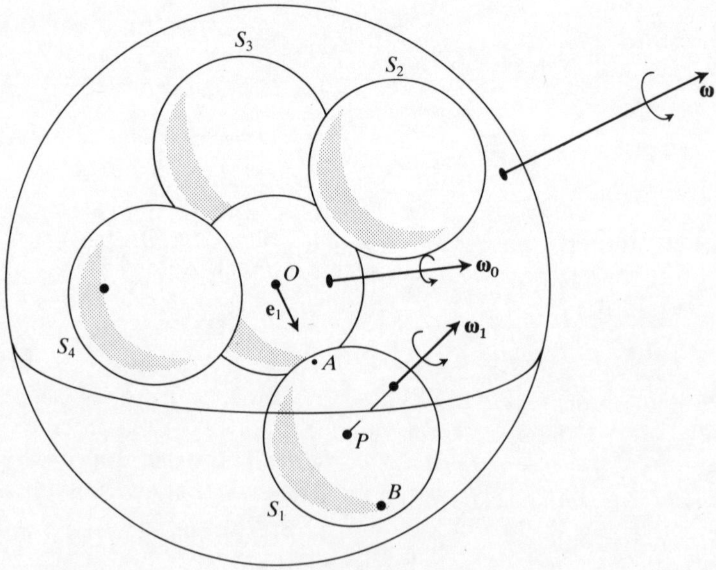

Fig. 1.4.3

(the dot is the time derivative with respect to $\hat{R}$), then the distances between the points are constant.

This follows from

$$\frac{\mathrm{d}}{\mathrm{d}t}[(\boldsymbol{r}_i-\boldsymbol{r}_j)\,.\,(\boldsymbol{r}_i-\boldsymbol{r}_j)] = 2[\boldsymbol{\theta}\wedge(\boldsymbol{r}_i-\boldsymbol{r}_j)]\,.\,(\boldsymbol{r}_i-\boldsymbol{r}_j)$$
$$=0. \tag{1.4.7}$$

Consider S_1. Let its angular velocity be $\boldsymbol{\omega}_1$; and let the angular velocities of S_0 and S be $\boldsymbol{\omega}_0$ and $\boldsymbol{\omega}$. Let $\boldsymbol{e}_1$ be the unit vector from O towards the centre P of S_1, so that the O-position vector of P is $\boldsymbol{r}_1 = 2a\boldsymbol{e}_1$.

The rolling condition at the point A of contact between S_1 and S_0 is

$$\boldsymbol{\omega}_0\wedge(a\boldsymbol{e}_1) = \dot{\boldsymbol{r}}_1 + \boldsymbol{\omega}_1\wedge(-a\boldsymbol{e}_1) \tag{1.4.8}$$

(the left-hand side is the velocity of the particle of S_0 at A; the right-hand side is the velocity of the particle of S_1 at A). Similarly, the rolling condition at the point B of contact between S_1 and S is

$$\boldsymbol{\omega}\wedge(3a\boldsymbol{e}_1) = \dot{\boldsymbol{r}}_1 + \boldsymbol{\omega}_1\wedge(a\boldsymbol{e}_1). \tag{1.4.9}$$

By adding

$$\dot{\boldsymbol{r}}_1 = \frac{a}{2}(3\boldsymbol{\omega}+\boldsymbol{\omega}_0)\wedge\boldsymbol{e}_1 = (\tfrac{3}{4}\boldsymbol{\omega}+\tfrac{1}{4}\boldsymbol{\omega}_0)\wedge\boldsymbol{r}_1. \tag{1.4.9}$$

By the same argument applied to S_2, S_3, and S_4,

$$\dot{\boldsymbol{r}}_i = \boldsymbol{\theta}\wedge\boldsymbol{r}_i, \qquad i=1,2,3,4, \tag{1.4.10}$$

where $\boldsymbol{\theta} = \frac{3}{4}\boldsymbol{\omega}+\frac{1}{4}\boldsymbol{\omega}_0$ and the $\boldsymbol{r}_i$ are the O-position vectors of the centres of S_1, S_2, S_3, and S_4. Therefore the distances between the centres remain constant and the result follows.

Exercises

(1.4.1) Establish the properties of the line L defined by eqn (1.4.4) for the general motion of a rigid body: that is, show that if P is a point of L, then $\boldsymbol{v}_P$ is proportional to $\boldsymbol{\omega}$; and show conversely that if $\boldsymbol{v}_P$ is proportional to $\boldsymbol{\omega}$, then P lies on L. (Assume that $\boldsymbol{\omega}\neq 0$.)

(1.4.2) A rigid body has a right-handed motion relative to a frame R. If the body is observed in a mirror, does the motion appear to be right or left-handed? If the motion is filmed and the film is then run backwards, does the motion appear to be right- or left-handed?

(1.4.3)† A rigid body has angular velocity $\boldsymbol{\omega}$ and has one point O fixed (relative to a frame $\hat{R}$). Show that if $\boldsymbol{\omega}\wedge\hat{\mathrm{D}}\boldsymbol{\omega}\neq 0$, then O is the only point with zero acceleration relative to $\hat{R}$.

(1.4.4) Show that the motion of a rigid body is determined at any instant by the velocities of three noncollinear points.

Three particles A, B, and C have velocities $\boldsymbol{u}$, $\boldsymbol{v}$, and $\boldsymbol{w}$ respectively relative to a frame $\hat{R}$. Show that they can belong to a rigid body if and only if

$$(\boldsymbol{a}-\boldsymbol{b}) \cdot (\boldsymbol{u}-\boldsymbol{v}) = (\boldsymbol{b}-\boldsymbol{c}) \cdot (\boldsymbol{v}-\boldsymbol{w}) = (\boldsymbol{c}-\boldsymbol{a}) \cdot (\boldsymbol{w}-\boldsymbol{u}) = 0$$

where $\boldsymbol{a}$, $\boldsymbol{b}$, and $\boldsymbol{c}$ are the position vectors of A, B, and C from the origin of $\hat{R}$.

(1.4.5) A sphere of radius a is rolling without slipping on a rough horizontal plane in such a way that its centre traces out a horizontal circle, radius b and centre O, with constant angular speed Ω. Let $(\boldsymbol{i}, \boldsymbol{j}, \boldsymbol{k})$ be an orthonormal triad with $\boldsymbol{k}$ vertical and $\boldsymbol{i}$ in the direction from O to the centre of the sphere. Show that the angular velocity of the sphere relative to the plane satisfies

$$\boldsymbol{\omega} = n\boldsymbol{k} - \frac{b}{a}\Omega \boldsymbol{i}$$

where $n = \boldsymbol{\omega} \cdot \boldsymbol{k}$. Show that if n is constant, then the locus of the point of contact on the sphere is a circle. What is its radius?

2 Lagrangian mechanics

2.1 An example

Our task in this chapter is to develop some techniques for handling constraints in mechanical systems and for finding and solving dynamical equations in general coordinates. We shall start with an example.

A bead of mass m is sliding along a smooth wire in the shape of the curve

$$\frac{x^2}{a^2}+\frac{y^2}{b^2}=1, \qquad z=c\cos^{-1}\left(\frac{x}{a}\right), \tag{2.1.1}$$

where $x=r_1$, $y=r_2$, and $z=r_3$ are Cartesian coordinates in an inertial frame $R=(O, B)$. The only forces on the bead are gravity, acting in the negative z-direction, and the normal reaction $\boldsymbol{N}$ of the wire (Fig. 2.1.1).

By Newton's second law

$$m\ddot{x}=N_1, \qquad m\ddot{y}=N_2, \qquad m\ddot{z}=-mg+N_3. \tag{2.1.2}$$

Here we have three equations in the six unknowns x, y, z, N_1, N_2, and N_3. To these we can add the two constraint equations (2.1.1) and the condition that $\boldsymbol{N}$ should be orthogonal to the velocity $\boldsymbol{v}$ of the bead (which is tangent to the wire):

$$\boldsymbol{v}\,.\,\boldsymbol{N}=\dot{x}N_1+\dot{y}N_2+\dot{z}N_3=0. \tag{2.1.3}$$

In all, six equations in six unknowns. In principle, therefore, the motion is determined; but, in practice, a frontal attack by differentiation and elimination might not lead very directly to the solution.

A more sensible approach is to write the equation of the wire in parametric form $\boldsymbol{r}=\boldsymbol{r}(q)$, by putting

$$x=a\cos q, \qquad y=b\sin q, \qquad z=cq. \tag{2.1.4}$$

Then the components of the acceleration relative to R are

$$\begin{aligned}\ddot{x}&=-a\dot{q}^2\cos q-a\ddot{q}\sin q,\\ \ddot{y}&=-b\dot{q}^2\sin q+b\ddot{q}\cos q,\\ \ddot{z}&=c\ddot{q}.\end{aligned} \tag{2.1.5}$$

By combining eqns (2.1.3) and (2.1.2) to get

$$m\dot{x}\ddot{x}+m\dot{y}\ddot{y}+m\dot{z}(\ddot{z}+g)=0 \tag{2.1.6}$$

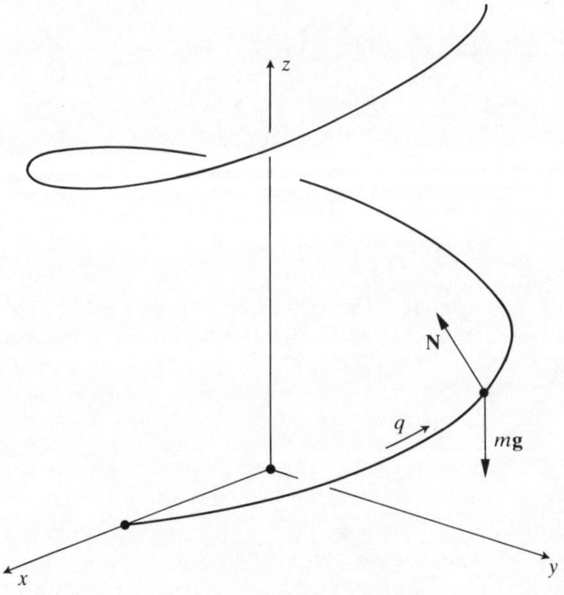

Fig. 2.1.1

and then substituting from eqn (2.1.5), we obtain

$$\ddot{q}(a^2 \sin^2 q + b^2 \cos^2 q + c^2) + \dot{q}^2(a^2 - b^2)\cos q \sin q + cg = 0, \quad (2.1.7)$$

after some cancellation and rearrangement. This is better: we now have a single second-order differential equation for the single unknown q. The irrelevant normal reaction has been eliminated. But we have had to undertake the unnecessary task of calculating the components of the acceleration in terms of the first two time derivatives of q; easy enough in this particular case, but a source of complication in more general problems.

In vector notation, eqn (2.1.2) is

$$\dot{\boldsymbol{p}} = \boldsymbol{N} - mg\boldsymbol{k} \quad (2.1.8)$$

where the dot is the time derivative with respect to B, $\boldsymbol{k}$ is the unit vector along the z-axis, and $\boldsymbol{p} = m\dot{\boldsymbol{r}}$ is the momentum. Equation (2.1.7) follows on taking the scalar product with $\mathrm{d}\boldsymbol{r}/\mathrm{d}q$, which is tangent to the wire and therefore orthogonal to $\boldsymbol{N}$.

Rather than go direct from eqn (2.1.8) to eqn (2.1.7), however, we can do the calculation in a slightly different way, which, at first sight, may seem a little obscure; but it is worth following carefully because it contains the central idea of the Lagrangian theory.

Put

$$p = \boldsymbol{p} \cdot \frac{\mathrm{d}\boldsymbol{r}}{\mathrm{d}q}, \qquad F = -mg\boldsymbol{k} \cdot \frac{\mathrm{d}\boldsymbol{r}}{\mathrm{d}q}. \tag{2.1.9}$$

We shall call p and F the *q-components* of the momentum and the gravitational force. Now

$$\dot{p} = \dot{\boldsymbol{p}} \cdot \frac{\mathrm{d}\boldsymbol{r}}{\mathrm{d}q} + \boldsymbol{p} \cdot \frac{\mathrm{d}^2\boldsymbol{r}}{\mathrm{d}q^2}\dot{q} = F + \boldsymbol{p} \cdot \frac{\mathrm{d}^2\boldsymbol{r}}{\mathrm{d}q^2}\dot{q} \tag{2.1.10}$$

(from eqn (2.1.8)). But for the 'correction term'

$$\boldsymbol{p} \cdot \frac{\mathrm{d}^2\boldsymbol{r}}{\mathrm{d}q^2}\dot{q} = m\frac{\mathrm{d}\boldsymbol{r}}{\mathrm{d}q} \cdot \frac{\mathrm{d}^2\boldsymbol{r}}{\mathrm{d}q^2}\dot{q}^2, \tag{2.1.11}$$

eqn (2.1.10) gives a simple and appealing equality between the time derivative of p and the q-component of the gravitational force.

The key fact that underlies the Lagrangian approach is that both p and the correction term can be obtained very simply from the expression for the kinetic energy; not just in this system, but, as we shall soon see, in a wide class of systems with an arbitrary number of degrees of freedom.

In our example the kinetic energy (relative to R) is

$$T = \tfrac{1}{2}m\dot{\boldsymbol{r}} \cdot \dot{\boldsymbol{r}} = \tfrac{1}{2}m\frac{\mathrm{d}\boldsymbol{r}}{\mathrm{d}q} \cdot \frac{\mathrm{d}\boldsymbol{r}}{\mathrm{d}q}v^2, \tag{2.1.12}$$

where $v = \dot{q}$ and $\boldsymbol{r}$ is expressed as a function of q through eqns (2.1.4); T is therefore a function of the two variables q and v, which together determine the position of the bead on the wire and its state of motion. Explicitly,

$$T = \tfrac{1}{2}m(a^2 \sin^2 q + b^2 \cos^2 q + c^2)v^2. \tag{2.1.13}$$

By taking the partial derivatives of T, first with respect to v with q held fixed and then with respect to q with v held fixed, we find that

$$\begin{aligned} \frac{\partial T}{\partial v} &= m\frac{\mathrm{d}\boldsymbol{r}}{\mathrm{d}q} \cdot \frac{\mathrm{d}\boldsymbol{r}}{\mathrm{d}q}v = p, \\ \frac{\partial T}{\partial q} &= m\frac{\mathrm{d}^2\boldsymbol{r}}{\mathrm{d}q^2} \cdot \frac{\mathrm{d}\boldsymbol{r}}{\mathrm{d}q}v^2. \end{aligned} \tag{2.1.14}$$

In other words, p is the partial derivative $\partial T/\partial v$ and the correction term is $\partial T/\partial q$. Therefore the equation of motion is

$$\frac{\mathrm{d}}{\mathrm{d}t}\left(\frac{\partial T}{\partial v}\right) - \frac{\partial T}{\partial q} = F. \tag{2.1.15}$$

Thus the work of finding the equation of motion is all but done once the kinetic energy has been expressed in terms of q and v. The gain that that represents will soon be obvious.

There is considerable scope for confusion in the interpretation of the various derivatives in eqn (2.1.15), and some attention will be given to their precise meaning when we come to derive the general version. For the moment, we must be content with the following recipe for decoding the left-hand side.

> (1) Express T in terms of q and v.
> (2) Take the partial derivatives $\partial T/\partial q$ and $\partial T/\partial v$ by treating q and v as if they were independent variables; in other words, forget that $v = \dot{q}$.
> (3) After finding the partial derivatives, substitute $\dot{q}$ for v and then take the time derivative of $\partial T/\partial v$, regarding q as a function of time.

In this case, the first step gives eqn (2.1.13) and the second gives

(2)
$$\frac{\partial T}{\partial q} = m(a^2 - b^2)v^2 \sin q \cos q,$$
$$\frac{\partial T}{\partial v} = m(a^2 \sin^2 q + b^2 \cos^2 q + c^2)v. \tag{2.1.16}$$

Hence the equation of motion is

(3)
$$\frac{\mathrm{d}}{\mathrm{d}t}[m(a^2 \sin^2 q + b^2 \cos^2 q + c^2)\dot{q}] - m(a^2 - b^2)\dot{q}^2 \sin q \cos q = -mcg, \tag{2.1.17}$$

which immediately reduces to eqn (2.1.7).

2.2 Configuration space and phase space

Consider a system made up of N particles with masses m_α ($\alpha = 1, 2, \ldots, N$), subject to forces $\boldsymbol{F}_\alpha$. In an inertial frame $R = (O, B)$, let the position vector of particle α be $\boldsymbol{r}_\alpha$ and let its coordinates be $(r_{\alpha 1}, r_{\alpha 2}, r_{\alpha 3})$. We shall allow the $\boldsymbol{F}_\alpha$ to depend on the positions and velocities of all the particles.

As the system evolves, each particle traces out a curve in space. Thus the motion of the system is represented by a set of N curves.

The first step in the Lagrangian theory is to make a shift in viewpoint:

we must think of these N curves as a single orbit in a $3N$-dimensional space. This is the *configuration space* C in which the points are labelled by the $3N$ coordinates

$$r_{11}, r_{12}, r_{13}, r_{21}, r_{22}, \ldots, r_{N3}. \tag{2.2.1}$$

Each point of C corresponds to a particular configuration of the system and as the particles move, their successive configurations trace out a curve in C.

To emphasize that C is to be thought of as a single space, rather than as a Cartesian product of N copies of three-dimensional Euclidean space, we relabel the coordinates, by writing

$$q_1 = r_{11}, \quad q_2 = r_{12}, \quad q_3 = r_{13}, \quad q_4 = r_{21}, \quad q_5 = r_{22}, \ldots, q_n = r_{N3}, \tag{2.2.2}$$

where $n = 3N$; and we introduce constants

$$\mu_1 = m_1, \quad \mu_2 = m_1, \quad \mu_3 = m_1, \quad \mu_4 = m_2, \quad \mu_5 = m_2, \ldots, \mu_n = m_N. \tag{2.2.3}$$

Then the equations of motion become

$$\mu_1 \ddot{q}_1 = F_1, \quad \mu_2 \ddot{q}_2 = F_2, \ldots, \mu_n \ddot{q}_n = F_n \tag{2.2.4}$$

where F_1, F_2, F_3 are the components of $\boldsymbol{F}_1$; F_4, F_5, F_6 are the components of $\boldsymbol{F}_2$; and so on. Each F_a $(a = 1, 2, \ldots, n)$ is a function of the coordinates $q_1, q_2, \ldots, q_n$ and the velocities $\dot{q}_1, \dot{q}_2, \ldots, \dot{q}_n$.

The solution to this system of differential equations depends on the initial values of q_a and $\dot{q}_a$ $(a = 1, 2, \ldots, n)$; so through each point of C, there is a different orbit for each choice of the velocities $\dot{q}_a$.

To get a cleaner picture, we double the number of variables, by introducing the *phase space* P in which

$$q_1, q_2, \ldots, q_n, v_1, v_2, \ldots, v_n \tag{2.2.5}$$

are coordinates and then write the equations of motion in the form

$$\begin{gathered} \mu_1 \dot{v}_1 = F_1, \ldots, \mu_n \dot{v}_n = F_n \\ \dot{q}_1 = v_1, \ldots, \dot{q}_n = v_n. \end{gathered} \tag{2.2.6}$$

This is a common trick in the theory of differential equations: we have replaced the n second-order differential equations (2.2.4) by $2n$ first-order equations (2.2.6).

Each point of P corresponds to a particular *state of motion*: a particular configuration together with a particular set of velocities for the particles.

The doubled-up equations of motion have a unique solution for each choice of the initial values of q_a and v_a. In P, therefore, there is just one orbit through each point.

It may be that we also want to consider forces that depend on time as well as on configuration and velocity. In this case the equations of motion are of the form

$$\mu\ddot{q} = F(q_1, \ldots, q_n, v_1, \ldots, v_n, t) \tag{2.2.7}$$

and it is convenient to picture the motion in the time-configuration space CT (which has coordinates $q_1, q_2, \ldots, q_n, t$) or the time-phase space PT (which has coordinates $q_1, \ldots, q_n, v_1, \ldots, v_n, t$). Again, there is just one orbit through each point of PT.

Similar ideas arise for systems in one and two dimensions.

The following examples show how one can use the pattern of the orbits in P to understand the dynamical behaviour of systems with one degree of freedom.

Example (2.2.1) The harmonic oscillator. A particle of mass m moves along the x-axis under the influence of the force $F = -mx$. There is just one equation of motion

$$m\ddot{x} = -mx. \tag{2.2.8}$$

The configuration space C is one-dimensional (with coordinate $q = x$) and the phase space P is two-dimensional (with coordinates q and v). The equations of motion in P are

$$\dot{v} = -q, \qquad \dot{q} = v \tag{2.2.9}$$

and (since these imply that $q^2 + v^2 = \text{const.}$) the orbits are circles.

Figure 2.2.1 illustrates the 'phase portrait' of the system. The arrows point in the direction of *increasing* q when $v = \dot{q} > 0$; and in the direction of *decreasing* q when $v < 0$. On each orbit, q oscillates between two equal

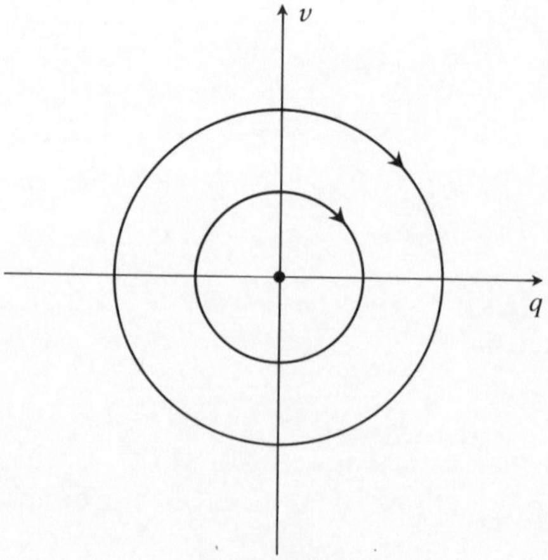

Fig. 2.2.1

and opposite extremes—the intersection points of the circle with the q-axis.

The only point at which the particle can remain at rest is the origin $q = 0$. This is a position of *stable* equilibrium since if the particle is displaced slightly, then its orbit in P is a small circle centre $(0, 0)$, on which q and v remain close to zero.

Example (2.2.2). If instead $F = mx$, then the orbits are the hyperbolas $v^2 - q^2 = \text{constant}$ and the origin is a position of *unstable* equilibrium. There are several different types of orbit, illustrated in Fig. 2.2.2.

(A) $v^2 - q^2 < 0$, $q > 0$: the particle approaches the origin from $q = \infty$; q comes to a minimum (the intersection point with the q-axis) and then increases again without ever reaching $q = 0$.

(B) $v^2 - q^2 = 0$, $q > 0$, $v > 0$: here $\dot{q} = q$ and the particle moves away from the equilibrium position with q increasing exponentially with time.

(C) $v^2 - q^2 = 0$, $q > 0$, $v < 0$: here $\dot{q} = -q$. The particle approaches the origin, with q decreasing like e^{-t}, but never reaches it.

(D) $v^2 - q^2 > 0$, $v > 0$: the particle approaches the origin from $q = -\infty$, slows down, passes through $q = 0$, and then accelerates again towards $q = \infty$.

Example (2.2.3). A more interesting case is $F = 2mx\,(1 - x^2)$, for which the orbits are

$$(q - 1)^2(q + 1)^2 + v^2 = \text{constant} \tag{2.2.10}$$

(Fig. 2.2.3). There are two positions of stable equilibrium at $q = -1$ and $q = 1$, near which the system behaves like the harmonic oscillator; and one position of unstable equilibrium at $q = 0$, near which it behaves like

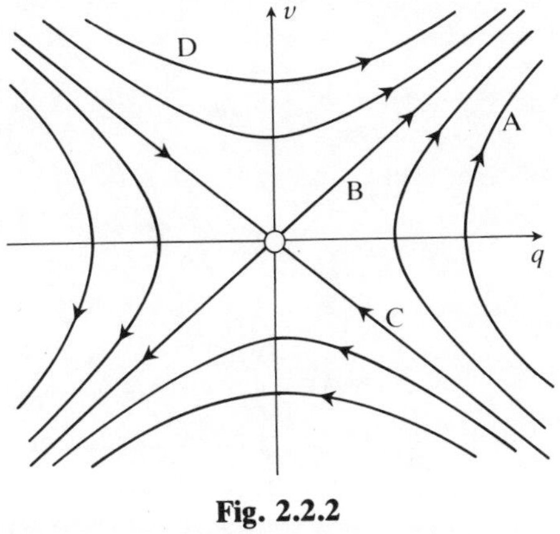

Fig. 2.2.2

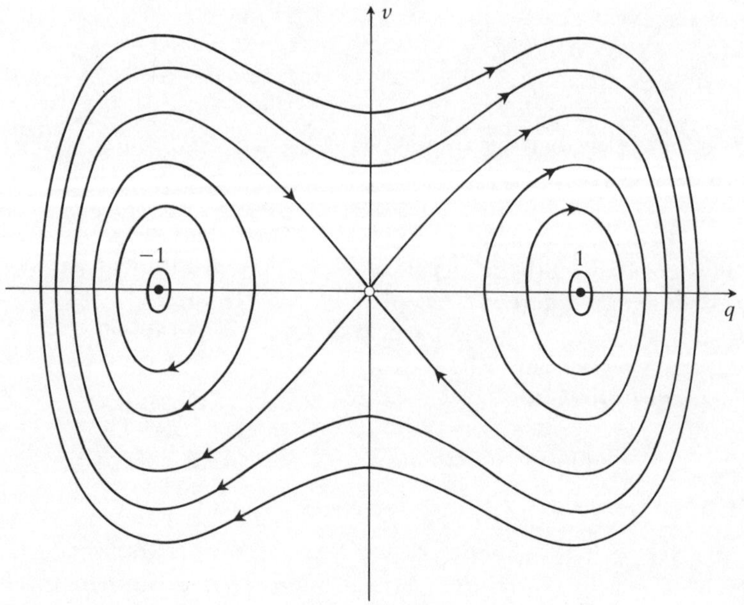

Fig. 2.2.3

the previous example (although the behaviour for larger values of $|q|$ is very different).

In general, if $F = F(x)$, then the equilibrium points are the values of x for which $F(x) = 0$: they are stable when $F'(x) < 0$ and unstable when $F'(x) > 0$. More details are given in Jordan and Smith's[6] *Nonlinear ordinary differential equations.* □

'Phase plane analysis' is a powerful tool for tackling nonlinear ordinary differential equations in a general context. As an illustration, we shall return to Foucault's pendulum.

Example (2.2.4) The area effect. (Continuation of example (1.3.3)). By dividing eqn (1.3.29) by $\dot{\theta}^2$, and substituting $h = r^2\dot{\theta}$ and $r^2 = a^2 - z^2$ (which implies that $r\dot{r} = -z\dot{z}$), we obtain

$$\tfrac{1}{2}v^2 = \frac{(a^2 - z^2)^2}{2a^2h^2}(F(z) - h^2), \tag{2.2.11}$$

where $v = \mathrm{d}z/\mathrm{d}\theta$ and

$$F(z) = 2(E - g'z)(a^2 - z^2). \tag{2.2.12}$$

Apart from the fact that z rather than q is the dependent variable and θ rather than t is the independent variable, this is the same sort of equation as those that arose in the previous examples.

As the pendulum oscillates, both v and z vary as functions of θ and so the different possible motions give rise to a family of curves in the z, v-plane (the 'phase plane').

Consider first the case $\Omega = 0$ (no rotation). Then $g' = g$, $\dot{h} = 0$, and the phase plane curves, which are given by eqn (2.2.11), are labelled by the various values of the constants E and h (we must exclude the singular case $h = 0$ in which the pendulum oscillates in a vertical plane and θ is not a good parameter).

Fix a value of $E > -ag$. Then there is a critical value h_0 of h for which the cubic $F(z) - h^2$ has a repeated root $z = z_0$. Since z_0 is also a root of $F'(z)$, we have

$$g(a^2 - z_0^2) + 2z_0(E - gz_0) = 0. \tag{2.2.13}$$

Hence

$$z_0 = \frac{1}{3g}(E - \surd(E^2 + 3a^2g^2)), \tag{2.2.14}$$

the sign of the square root being chosen by sketching the graph of $F(z)$ (Fig. 2.2.4).

The phase plane orbit labelled by $h = h_0$ is the single point $(z_0, 0)$: here the path of the bob is a horizontal circle in the plane $z = z_0$.

We shall look at what happens when the motion is close to this horizontal circle. The argument involves two approximations: we assume that the radius of the circle is small compared with a (that is $a - |z_0| \ll a$); and that the departure of the path of the bob from the circle is small compared with the radius of the circle (that is, $|z - z_0| \ll a - |z_0|$).

If h is near the critical value h_0, so that $0 < |h_0 - h| \ll |h_0|$, we have $z = z_0 + y$ where $|y| \ll a - |z_0|$; and, by ignoring terms of order y^3, we find from eqn (2.2.11) that the dependence of y on θ is determined by

$$\frac{1}{2}\left(\frac{dy}{d\theta}\right)^2 = \frac{(a^2 - z_0^2)^2}{2a^2h_0^2}(h_0^2 - h^2 + \tfrac{1}{2}y^2F''(z_0)) \tag{2.2.15}$$

(this requires a little thought: the crucial point is that $h_0^2 - h^2$ is of the same order as y^2). On differentiating with respect to y,

$$\frac{d^2y}{d\theta^2} + \lambda^2 y = 0 \tag{2.2.16}$$

where

$$\lambda^2 = -\frac{(a^2 - z_0^2)^2F''(z_0)}{2a^2h_0^2} = \frac{2z_0(3gz_0 - E)}{a^2g} \tag{2.2.17}$$

(we have used eqn (2.2.13) together with $h_0^2 = F(z_0)$).

Therefore y, and hence also z, oscillate as functions of θ with period $2\pi/\lambda$: the bob moves close to a horizontal circle, but it rises and falls slightly as it rotates about the vertical axis.

If λ were an integer, then we should have $z(0) = z(2\pi)$ and the bob would return to its initial position after each complete circuit. This is what happens in the limit $E \to -ag$, $z_0 \to -a$, when the circular path collapses to a point: then eqn (2.2.17) gives $\lambda^2 = 4$.

But if E is slightly above the limiting value, so that $E = -ag + \varepsilon g$ where

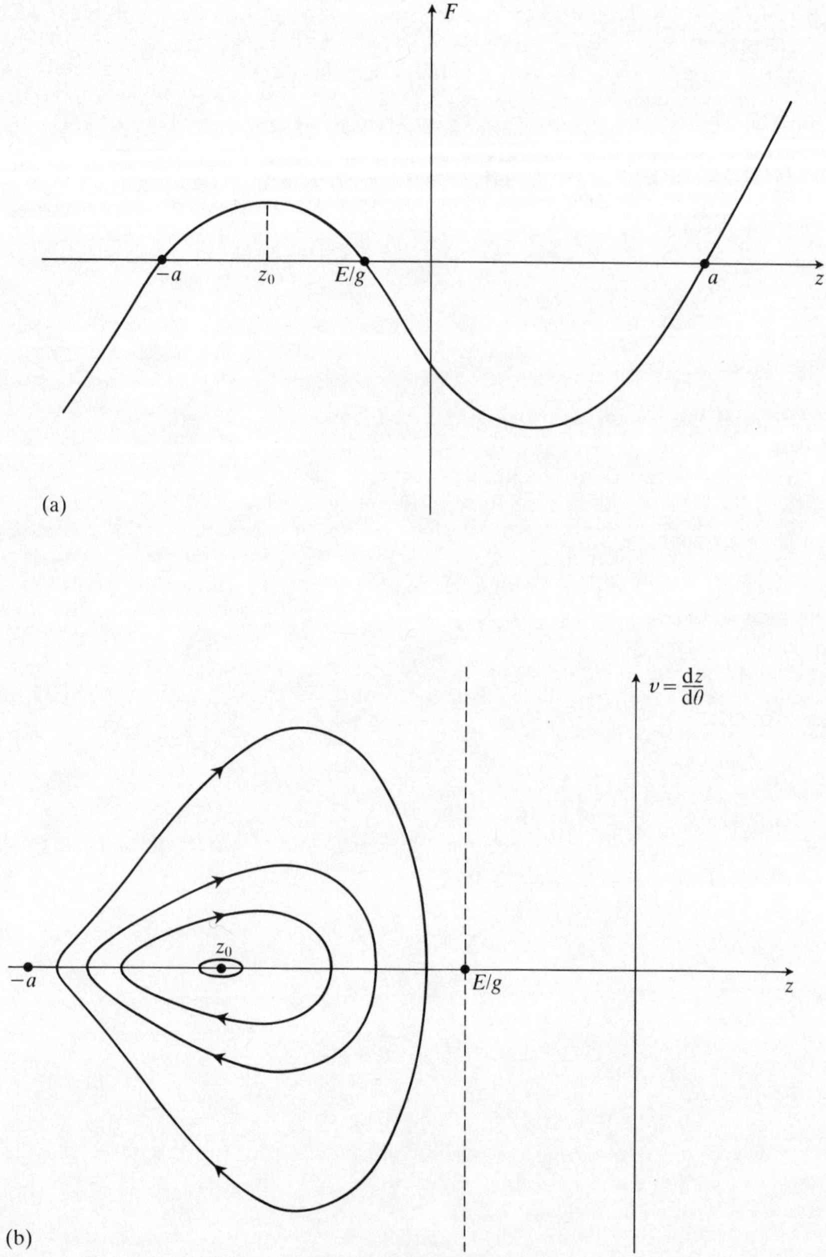

Fig. 2.2.4 (a) The graph $F(z)$. (b) All the orbits in the z, v-plane lie to the left of $z = E/g$ (the broken line).

$|y| \ll \varepsilon \ll a$, then

$$z_0 = -a + \tfrac{1}{2}\varepsilon + O(\varepsilon^2/a^2) \tag{2.2.18}$$

(from eqn (2.2.13)) and

$$\lambda^2 = 4(1 - 3\varepsilon/4a) + O(\varepsilon^2/a^2) \tag{2.2.19}$$

(from eqn (2.2.17)). Thus the period of z as a function of θ is

$$\pi\left(1 + \frac{3\varepsilon}{8a}\right) = \pi + \frac{3A}{8a^2} \tag{2.2.20}$$

where A is the area of the circular orbit. On each complete revolution, the point at which z reaches its maximum advances through an angle $3A/4a^2$, in the same sense as the bob's motion. This is the *area effect.*

The derivation is based on the assumption that $|y| \ll \varepsilon \ll a$. A more careful analysis shows that the area effect is given by the same expression when y is comparable with ε: then the projection of the path of the bob into the horizontal r, θ-plane is approximately an ellipse with its centre at the origin, but the axes of the ellipse rotate through an angle $3A/4a^2$ in the forward sense on each revolution, where A is the area of the ellipse.

When $\Omega \neq 0$, the area effect is superimposed on the Coriolis rotation and dominates it unless the pendulum is set in motion in such a way that $A^2 g \ll \Omega^2 a^5$. □

Exercises

(2.2.1) Sketch the orbits of the harmonic oscillator in CT (coordinates q and t) and PT (coordinates q, v, and t).

(2.2.2) Sketch the orbits in P for a particle of mass m moving along the x-axis under the influence of
(i) the force $m(\frac{3}{2}x^2 - x)$,
(ii) the force $-m \sin x$.
In both cases describe the various motions.

2.3 Coordinate transformations

We shall now consider the dynamics of our system of particles in a general coordinate system. We shall introduce new coordinates $\tilde{q}_a$ and $\tilde{t}$ on CT which are related to q_a and t by expressions of the form

$$\tilde{q}_a = \tilde{q}_a(q_1, q_2, \ldots, q_n, t), \qquad \tilde{t} = t. \tag{2.3.1}$$

For a single particle, the $\tilde{q}_a$ might be Cartesian coordinates in a rotating frame, or spherical polar coordinates. For two particles, they might be the three coordinates of the centre of mass together with three other coordinates that determine the position of one of the particles relative to the centre of mass; and so on.

The first step is to extend the coordinate transformation to PT.

A curve γ in CT, such as an orbit of the system, can be specified by giving the q_a as functions of time:

$$q_a = q_a(t). \tag{2.3.2}$$

Definition (2.3.1). The *extension of* γ is the curve Γ in PT given by $q_a = q_a(t)$, $v_a = \dot{q}_a(t)$.

Example (2.3.1). For a system with one degree of freedom, the extension of $q = \cos t$ is the helix $q = \cos t$, $v = -\sin t$ (Fig. 2.3.1). □

In the new coordinates, eqn (2.3.2) becomes $\tilde{q}_a = \tilde{q}_a(t)$; and it is natural to introduce new velocity coordinates $\tilde{v}_a$ so that the extension to PT is given by $\tilde{q}_a = \tilde{q}_a(t)$, $\tilde{v}_a = \dot{\tilde{q}}_a(t)$.

By applying the chain rule,

$$\dot{\tilde{q}}_a(t) = \frac{\partial \tilde{q}_a}{\partial q_1}\dot{q}_1 + \frac{\partial \tilde{q}_a}{\partial q_2}\dot{q}_2 + \cdots + \frac{\partial \tilde{q}_a}{\partial q_n}\dot{q}_n + \frac{\partial \tilde{q}_a}{\partial t} \tag{2.3.3}$$

for $a = 1, 2, \ldots, n$. In PT, therefore, we have a coordinate transformation of the form

$$\begin{aligned} \tilde{q}_a &= \tilde{q}_a(q, t), \\ \tilde{v}_a &= \tilde{v}_a(q, v, t) = \frac{\partial \tilde{q}_a}{\partial q_b} v_b + \frac{\partial \tilde{q}_a}{\partial t}, \\ \tilde{t} &= t. \end{aligned} \tag{2.3.4}$$

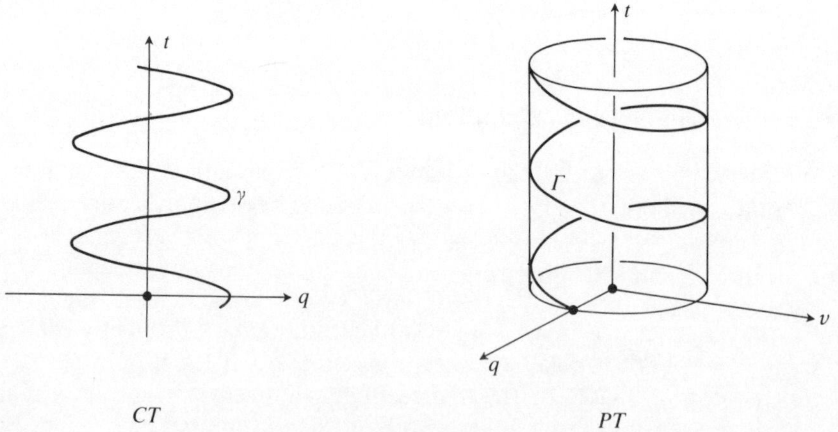

Fig. 2.3.1 The extension of $q = \cos t$ from time-configuration space CT to time-phase space PT.

Some explanation of the notation is needed: first, we are applying the *Einstein conventions* to the indices $a, b, c, \ldots = 1, 2, \ldots, n$. For an index in the first part of the alphabet, a repetition of the index in a term implies a sum over $1, 2, \ldots, n$; and any equation is understood to hold for all possible values of its free (unsummed) indices. Thus the second of eqns (2.3.4) holds for $a = 1, 2, \ldots, n$; and there is a sum over $b = 1, 2, \ldots, n$ in the expression $(\partial\tilde{q}_a/\partial q_b)v_b$.

Second, $\tilde{q}_a = \tilde{q}_a(q, t)$ is a convenient shorthand for $\tilde{q}_a = \tilde{q}_a(q_1, \ldots, q_n, t)$; and so on.

It is necessary that the coordinate transformation should be non-singular. In other words, we should be able to invert eqn (2.3.4) and express the original coordinates q_a, v_a, t as smooth functions of $\tilde{q}_a$, $\tilde{v}_a$, and $\tilde{t}$. The condition for this to be possible (at least locally) is given by the *inverse function theorem*. To invert eqn (2.3.1) and so obtain q_a as a function of $\tilde{q}_1, \ldots, \tilde{q}_n$ and $\tilde{t} = t$, we need the matrix with entries $\partial\tilde{q}_a/\partial q_b$ to be non-singular. When this condition holds, we can write $q_a = q_a(\tilde{q}, \tilde{t})$ and invert eqn (2.3.4) to obtain

$$v_a = \frac{\partial q_a}{\partial \tilde{q}_b}\tilde{v}_b + \frac{\partial q_a}{\partial \tilde{t}}. \tag{2.3.5}$$

Our task is to rewrite the equation of motion in terms of the new coordinates. The direct substitution of eqn (2.3.4) into (2.2.6), however, can lead to cumbersome and unmanageable expressions. A more elegant approach is to make use of the following fundamental lemma.

Proposition (2.3.1). Let γ be a curve in CT given by $q_a = q_a(t)$ and let $F: PT \to \mathbb{R}$. That is, F is a function of the $2n + 1$ variables q_a, v_a, and t. Then

$$\frac{\mathrm{d}}{\mathrm{d}t}\left(\frac{\partial F}{\partial v_a}\right) - \frac{\partial F}{\partial q_a} = \frac{\partial \tilde{q}_b}{\partial q_a}\left[\frac{\mathrm{d}}{\mathrm{d}t}\left(\frac{\partial F}{\partial \tilde{v}_b}\right) - \frac{\partial F}{\partial \tilde{q}_b}\right] \tag{2.3.6}$$

where $\mathrm{d}/\mathrm{d}t$ denotes differentiation along the extension of γ to PT.

Remember that on the right-hand side, there is a sum over $b = 1, 2, \ldots, n$.

Before turning to the proof, we must understand the meaning of the various derivatives that appear. The first point is that there is a distinction between $\mathrm{d}/\mathrm{d}t$ and $\partial/\partial t$. If $f = f(q, v, t)$ is a function on PT, then

(i) $\mathrm{d}f/\mathrm{d}t$ is obtained by substituting $q_a = q_a(t)$, $v_a = \dot{q}_a(t)$ into $f = f(q, v, t)$ and *then* differentiating with respect to t;

(ii) $\partial f/\partial t$ is obtained by differentiating f with respect to t, holding the values of q_a and v_a fixed.

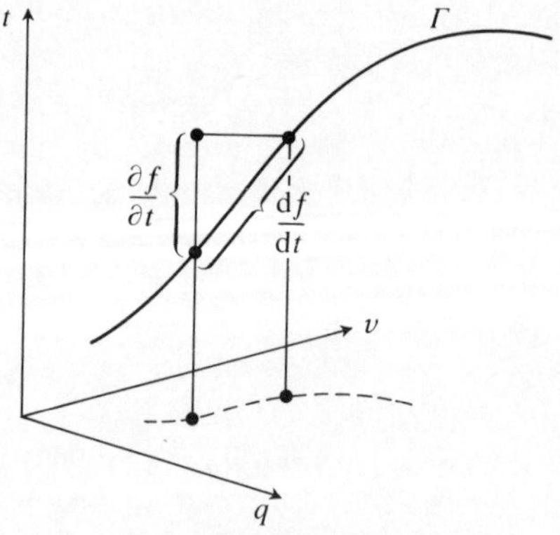

Fig. 2.3.2

The difference is illustrated in Fig. 2.3.2: when calculating $\mathrm{d}f/\mathrm{d}t$, one compares the values of f at two nearby points on the extended curve; for $\partial f/\partial t$, on the other hand, one compares the values of f at two different times, but for the same values of q_a and v_a (that is, for the same configuration and state of motion).

Example (2.3.2). With $n = 1$, $q = q_1$, and $v = v_1$: take $f = qvt$ and let γ be the curve $q = \cos t$. Then

$$\frac{\partial f}{\partial t} = qv = -\tfrac{1}{2}\sin 2t,$$
$$\frac{\mathrm{d}f}{\mathrm{d}t} = \frac{\mathrm{d}}{\mathrm{d}t}(-t\cos t\sin t) = -t\cos 2t - \tfrac{1}{2}\sin 2t. \tag{2.3.7}$$

□

The second point is that even though $t = \tilde{t}$, $\partial/\partial t$ is not necessarily the same as $\partial/\partial\tilde{t}$ (however, $\mathrm{d}/\mathrm{d}\tilde{t} = \mathrm{d}/\mathrm{d}t$; why?). This is an example of the *second fundamental confusion of calculus*, of which a simple demonstration is the following. Suppose that $g = g(x, y)$ and that $\tilde{x} = x$ and $\tilde{y} = x + y$. Then, by using the chain rule,

$$\frac{\partial g}{\partial x} = \frac{\partial g}{\partial \tilde{x}}\frac{\partial \tilde{x}}{\partial x} + \frac{\partial g}{\partial \tilde{y}}\frac{\partial \tilde{y}}{\partial x} = \frac{\partial g}{\partial \tilde{x}} + \frac{\partial g}{\partial \tilde{y}}, \tag{2.3.8}$$

so that $\partial g/\partial x \neq \partial g/\partial\tilde{x}$, even though $x = \tilde{x}$. The difficulty is that there is a potential ambiguity in the notation $\partial g/\partial x$: it does not make sense unless one knows what other variables are to be held fixed. In eqn (2.3.8), we held y fixed when we took the partial deriviative with respect to x; and $\tilde{y}$ fixed when we differentiated with respect to $\tilde{x}$.

Here and below we shall avoid the pitfalls of the second fundamental confusion by adopting the following device: when taking one of the partial derivatives $\partial/\partial\tilde{q}_a$, $\partial/\partial\tilde{v}_a$, or $\partial/\partial\tilde{t}$, we hold fixed all the other variables *with* tildes; but when taking $\partial/\partial q_a$, $\partial/\partial v_a$, or $\partial/\partial t$, we hold fixed the other variables *without* tildes. This allows us to distinguish between $\partial/\partial t$ and $\partial/\partial\tilde{t}$ in spite of the fact that $t=\tilde{t}$.

We shall go through the proof of proposition (2.3.1) first for the case of one degree of freedom (when there is one configuration coordinate $q=q_1$ and one velocity coordinate $v=v_1$) and then for the general case. The arguments are identical, but the details are easier to follow in the first case. An alternative proof will be given in section 2.5.

Proof (One degree of freedom). Here $F=F(q, v, t)$ and the coordinate transformation is

$$\begin{aligned}\tilde{q}&=\tilde{q}(q, t),\\ \tilde{v}&=\tilde{v}(q, v, t)=\frac{\partial\tilde{q}}{\partial q}v+\frac{\partial\tilde{q}}{\partial t},\\ \tilde{t}&=t.\end{aligned}\tag{2.3.9}$$

Therefore, by the chain rule,

$$\begin{aligned}\frac{\partial F}{\partial q}&=\frac{\partial F}{\partial\tilde{q}}\frac{\partial\tilde{q}}{\partial q}+\frac{\partial F}{\partial\tilde{v}}\frac{\partial\tilde{v}}{\partial q}+\frac{\partial F}{\partial\tilde{t}}\frac{\partial\tilde{t}}{\partial q}\\ &=\frac{\partial F}{\partial\tilde{q}}\frac{\partial\tilde{q}}{\partial q}+\frac{\partial F}{\partial\tilde{v}}\left[\frac{\partial^2\tilde{q}}{\partial q^2}v+\frac{\partial^2\tilde{q}}{\partial q\,\partial t}\right]\end{aligned}\tag{2.3.10}$$

since $\partial\tilde{t}/\partial q=0$ and $\partial\tilde{v}/\partial q$ is equal to the expression in square brackets (by differentiating the second of eqns (2.3.9)). Similarly

$$\begin{aligned}\frac{\partial F}{\partial v}&=\frac{\partial F}{\partial\tilde{q}}\frac{\partial\tilde{q}}{\partial v}+\frac{\partial F}{\partial\tilde{v}}\frac{\partial\tilde{v}}{\partial v}+\frac{\partial F}{\partial\tilde{t}}\frac{\partial\tilde{t}}{\partial v}\\ &=\frac{\partial F}{\partial\tilde{v}}\frac{\partial\tilde{q}}{\partial q}\end{aligned}\tag{2.3.11}$$

since $\partial\tilde{q}/\partial v=\partial\tilde{t}/\partial v=0$ and $\partial\tilde{v}/\partial v=\partial\tilde{q}/\partial q$, from eqn (2.3.9).

Now for any function $k=k(q, t)$ which depends only on q and t,

$$\frac{\mathrm{d}k}{\mathrm{d}t}=\frac{\partial k}{\partial q}\frac{\mathrm{d}q}{\mathrm{d}t}+\frac{\partial k}{\partial t}=\frac{\partial k}{\partial q}v+\frac{\partial k}{\partial t}.\tag{2.3.12}$$

Therefore

$$\frac{\mathrm{d}}{\mathrm{d}t}\left(\frac{\partial\tilde{q}}{\partial q}\right)=\frac{\partial^2\tilde{q}}{\partial q^2}v+\frac{\partial^2\tilde{q}}{\partial q\,\partial t},\tag{2.3.13}$$

and hence

$$\frac{\mathrm{d}}{\mathrm{d}t}\left(\frac{\partial F}{\partial v}\right)=\frac{\mathrm{d}}{\mathrm{d}t}\left(\frac{\partial F}{\partial \tilde{v}}\right)\frac{\partial \tilde{q}}{\partial q}+\frac{\partial F}{\partial \tilde{v}}\left[\frac{\partial^2 \tilde{q}}{\partial q^2}v+\frac{\partial^2 \tilde{q}}{\partial q\,\partial t}\right]. \tag{2.3.14}$$

The proposition now follows on subtracting eqn (2.3.10) from eqn (2.3.14). □

Proof (*n* degrees of freedom). The only difference is that when calculating partial derivatives by the chain rule, we must sum over the subscripts on the q_a's and v_a's. Thus

$$\begin{aligned}\frac{\partial F}{\partial q_a}&=\frac{\partial F}{\partial \tilde{q}_b}\frac{\partial \tilde{q}_b}{\partial q_a}+\frac{\partial F}{\partial \tilde{v}_b}\frac{\partial \tilde{v}_b}{\partial q_a}+\frac{\partial F}{\partial \tilde{t}}\frac{\partial \tilde{t}}{\partial q_a}\\&=\frac{\partial F}{\partial \tilde{q}_b}\frac{\partial \tilde{q}_b}{\partial q_a}+\frac{\partial F}{\partial \tilde{v}_b}\left[\frac{\partial^2 \tilde{q}_b}{\partial q_a\,\partial q_c}v_c+\frac{\partial^2 \tilde{q}_b}{\partial q_a\,\partial t}\right],\end{aligned} \tag{2.3.10'}$$

$$\begin{aligned}\frac{\partial F}{\partial v_a}&=\frac{\partial F}{\partial \tilde{q}_b}\frac{\partial \tilde{q}_b}{\partial v_a}+\frac{\partial F}{\partial \tilde{v}_b}\frac{\partial \tilde{v}_b}{\partial v_a}+\frac{\partial F}{\partial \tilde{t}}\frac{\partial \tilde{t}}{\partial v_a}\\&=\frac{\partial F}{\partial \tilde{v}_b}\frac{\partial \tilde{q}_b}{\partial q_a}.\end{aligned} \tag{2.3.11'}$$

Hence

$$\frac{\mathrm{d}}{\mathrm{d}t}\left(\frac{\partial F}{\partial v_a}\right)=\frac{\mathrm{d}}{\mathrm{d}t}\left(\frac{\partial F}{\partial \tilde{v}_b}\right)\frac{\partial \tilde{q}_b}{\partial q_a}+\frac{\partial F}{\partial \tilde{v}_b}\left[\frac{\partial^2 \tilde{q}_b}{\partial q_a\,\partial q_c}v_c+\frac{\partial^2 \tilde{q}_b}{\partial q_a\,\partial t}\right] \tag{2.3.14'}$$

since

$$\frac{\mathrm{d}}{\mathrm{d}t}\left(\frac{\partial \tilde{q}_b}{\partial q_a}\right)=\frac{\partial^2 \tilde{q}_b}{\partial q_a\,\partial q_c}v_c+\frac{\partial^2 \tilde{q}_b}{\partial q_a\,\partial t}. \tag{2.3.13'}$$

(Remember that the repetition of an index in a term implies a sum over $1, 2, \ldots, n$.) As before, the proposition (2.3.1) follows on subtracting (2.3.10′) from (2.3.14′). □

The proposition shows that the combination of derivatives

$$\frac{\mathrm{d}}{\mathrm{d}t}\left(\frac{\partial F}{\partial v_a}\right)-\frac{\partial F}{\partial q_a} \tag{2.3.15}$$

has a very straightforward transformation rule under a change of coordinates, although, as is clear from the proof, neither of the individual terms $\mathrm{d}/\mathrm{d}t(\partial F/\partial v_a)$ and $\partial F/\partial q_a$ behaves in a particularly simple way.

We are now in a position to transform the equations of motion. Let T be the total kinetic energy. Then

$$T=\tfrac{1}{2}(\mu_1 v_1^2+\mu_2 v_2^2+\cdots+\mu_n v_n^2), \tag{2.3.16}$$

where the μ_a's are defined by eqn (2.2.3). If we think of T as a function on PT, then

$$\frac{\partial T}{\partial v_1} = \mu_1 v_1, \qquad \frac{\partial T}{\partial v_2} = \mu_2 v_2, \ldots, \frac{\partial T}{\partial v_n} = \mu_n v_n. \tag{2.3.17}$$

Now $\partial T/\partial q_a = 0$, so the equations of motion can be written

$$\frac{\mathrm{d}}{\mathrm{d}t}\left(\frac{\partial T}{\partial v_a}\right) - \frac{\partial T}{\partial q_a} = F_a, \qquad \frac{\mathrm{d}}{\mathrm{d}t}(q_a) = v_a. \tag{2.3.18}$$

The point of this apparent complication is that the transformation to the new coordinates is now very straightforward. Apply the proposition to the first equation (with T replacing F)

$$\frac{\partial \tilde{q}_b}{\partial q_a}\left[\frac{\mathrm{d}}{\mathrm{d}t}\left(\frac{\partial T}{\partial \tilde{v}_b}\right) - \frac{\partial T}{\partial \tilde{q}_b}\right] = F_a. \tag{2.3.19}$$

Multiply both sides by $\partial q_a/\partial \tilde{q}_c$, sum over a, and make use of

$$\frac{\partial \tilde{q}_b}{\partial q_a}\frac{\partial q_a}{\partial \tilde{q}_c} = \frac{\partial \tilde{q}_b}{\partial \tilde{q}_c} = \delta_{bc} \tag{2.3.20}$$

(δ_{bc} is the Kronecker delta: it is equal to one when $b = c$ and to zero otherwise). The result is

$$\delta_{bc}\left[\frac{\mathrm{d}}{\mathrm{d}t}\left(\frac{\partial T}{\partial \tilde{v}_b}\right) - \frac{\partial T}{\partial \tilde{q}_b}\right] = \frac{\mathrm{d}}{\mathrm{d}t}\left(\frac{\partial T}{\partial \tilde{v}_c}\right) - \frac{\partial T}{\partial \tilde{q}_c} = \tilde{F}_c, \tag{2.3.21}$$

where the $\tilde{F}_c$ are defined by

$$\tilde{F}_c = \frac{\partial q_a}{\partial \tilde{q}_c} F_a \quad \text{or by} \quad F_a = \frac{\partial \tilde{q}_c}{\partial q_a}\tilde{F}_c. \tag{2.3.22}$$

Hence the equations of motion in the new coordinates are

$$\frac{\mathrm{d}}{\mathrm{d}t}\left(\frac{\partial T}{\partial \tilde{v}_a}\right) - \frac{\partial T}{\partial \tilde{q}_a} = \tilde{F}_a, \qquad \frac{\mathrm{d}}{\mathrm{d}t}(\tilde{q}_a) = \tilde{v}_a. \tag{2.3.23}$$

Definition (2.3.2). The $\tilde{F}_a$ are *generalized forces*, the $\tilde{q}_a$ are *generalized coordinates*, and the $\tilde{v}_a$ are *generalized velocities*.

If $q'_a = q'_a(q, t)$, $t' = t$ is an alternative system of generalized coordinates, then we can write $q'_a = q'_a(\tilde{q}, \tilde{t})$; and, by a calculation similar to that in eqn (2.3.3), the corresponding velocity coordinates are related by

$$v'_a = \frac{\partial q'_a}{\partial \tilde{q}_b}\tilde{v}_b + \frac{\partial q'_a}{\partial \tilde{t}}. \tag{2.3.24}$$

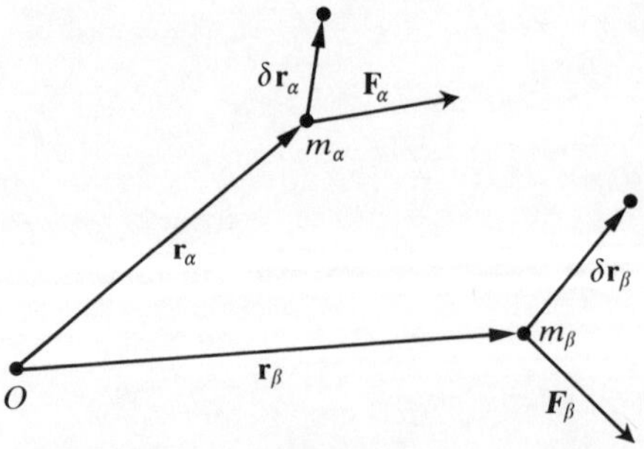

Fig. 2.3.3

The qualification 'generalized' will often be dropped. The $\tilde{F}_a$'s and $\tilde{v}_a$'s will also be referred to as the $\tilde{q}$-*components* of force and velocity.

If we know the values of the $\tilde{q}_a$ and $\tilde{t}$, then we also know the q_a and hence the positions of all the particles. Therefore $\boldsymbol{r}_\alpha = \boldsymbol{r}_\alpha(\tilde{q}, \tilde{t})$ and we can rewrite the definition of the generalized forces in the form

$$\tilde{F}_a = \sum_{\alpha=1}^{N} \boldsymbol{F}_\alpha \cdot \frac{\partial \boldsymbol{r}_\alpha}{\partial \tilde{q}_a}. \tag{2.3.25}$$

Suppose now that we make a small displacement in the configuration of the system by sending $\tilde{q}_a$ to $\tilde{q}_a + \delta\tilde{q}_a$, keeping $\tilde{t}$ fixed (Fig. 2.3.3). Then particle α is moved from the point with position vector $\boldsymbol{r}_\alpha + \delta\boldsymbol{r}_\alpha$, where

$$\delta\boldsymbol{r}_\alpha = \frac{\partial \boldsymbol{r}_\alpha}{\partial \tilde{q}_a} \delta\tilde{q}_a \tag{2.3.26}$$

(to the first order). Therefore

$$\tilde{F}_a \delta\tilde{q}_a = \sum_\alpha \boldsymbol{F}_\alpha \cdot \delta\boldsymbol{r}_\alpha. \tag{2.3.27}$$

The quantity on the right-hand side is the work done by the forces $\boldsymbol{F}_\alpha$ during the displacement.

This observation gives a convenient method for calculating the $\tilde{F}_a$'s: to find, for example, $\tilde{F}_1$, consider the displacement of the system given $\tilde{q}_1 \mapsto \tilde{q}_1 + \varepsilon$, with $\tilde{q}_2, \ldots, \tilde{q}_n$, and $\tilde{t}$ held fixed. Calculate (to the first order in ε) the total work done by all the forces during this displacement and equate the result to $\varepsilon\tilde{F}_1$.

Example (2.3.3). A particle of mass m moves under the influence of a force $\boldsymbol{F}$. The original coordinates are

$$q_1 = x, \qquad q_2 = y, \qquad q_3 = z, \tag{2.3.28}$$

where x, y, and z are Cartesian coordinates in some inertial frame. Take the $\tilde{q}_a$ to be spherical polar coordinates: $\tilde{q}_1 = r$, $\tilde{q}_2 = \theta$, $\tilde{q}_3 = \varphi$ where

$$\begin{aligned} x &= r \sin\theta \cos\varphi \\ y &= r \sin\theta \sin\varphi \\ z &= r \cos\theta. \end{aligned} \tag{2.3.29}$$

In terms of these

$$\begin{aligned} T &= \tfrac{1}{2}m(\dot{r}^2 + r^2\dot{\theta}^2 + r^2\dot{\varphi}^2 \sin^2\theta) \\ &= \tfrac{1}{2}m(\tilde{v}_1^2 + \tilde{q}_1^2\tilde{v}_2^2 + \tilde{q}_1^2\tilde{v}_3^2 \sin^2\tilde{q}_2). \end{aligned} \tag{2.3.30}$$

Thus the equations of motion are

$$\begin{aligned} \frac{\mathrm{d}}{\mathrm{d}t}\left(\frac{\partial T}{\partial \tilde{v}_1}\right) - \frac{\partial T}{\partial \tilde{q}_1} &= m\ddot{r} - mr(\dot{\theta}^2 + \dot{\varphi}^2 \sin^2\theta) = \tilde{F}_1 \\ \frac{\mathrm{d}}{\mathrm{d}t}\left(\frac{\partial T}{\partial \tilde{v}_2}\right) - \frac{\partial T}{\partial \tilde{q}_2} &= mr^2\ddot{\theta} + 2mr\dot{r}\dot{\theta} - mr^2\dot{\varphi}^2 \sin\theta\cos\theta = \tilde{F}_2 \\ \frac{\mathrm{d}}{\mathrm{d}t}\left(\frac{\partial T}{\partial \tilde{v}_3}\right) - \frac{\partial T}{\partial \tilde{q}_3} &= \frac{\mathrm{d}}{\mathrm{d}t}(mr^2\dot{\varphi}\sin^2\theta) = \tilde{F}_3. \end{aligned} \tag{2.3.31}$$

To find the $\tilde{q}$-components of $\boldsymbol{F}$, we follow the instructions: for example, for the φ-component $\tilde{F}_3$, we consider the displacement $\varphi \mapsto \varphi + \varepsilon$, with r, θ, and t fixed (Fig. 2.3.4). This changes the position of the particle by

$$\delta x = -\varepsilon r \sin\theta \sin\varphi, \qquad \delta y = \varepsilon r \sin\theta \cos\varphi, \qquad \delta z = 0 \tag{2.3.32}$$

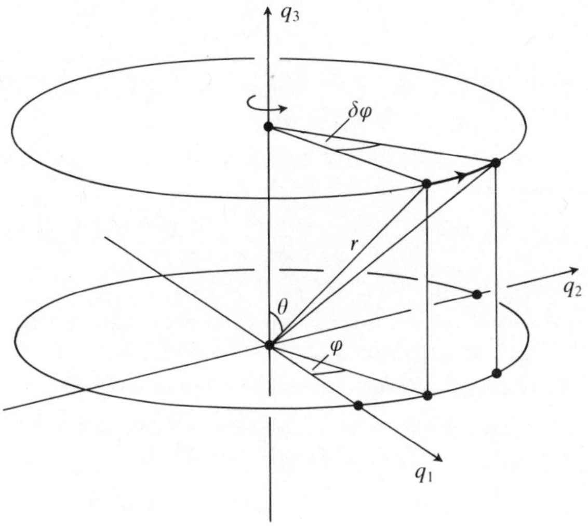

Fig. 2.3.4

to the first order in ε. Therefore, by cancelling ε,

$$\tilde{F}_3 = -F_1 r \sin\theta \sin\varphi + F_2 r \sin\theta \cos\varphi, \tag{2.3.33}$$

where F_1, F_2, and F_3 are the x, y, and z components of $\boldsymbol{F}$. □

There is a notable simplification in the case that the forces acting on the system are *conservative* in the sense that there exists a *potential.* That is a function $U = U(q, t)$ such that

$$F_a = -\frac{\partial U}{\partial q_a}. \tag{2.3.34}$$

Then the $\tilde{F}_a$ are given by

$$\tilde{F}_a = -\frac{\partial U}{\partial q_b}\frac{\partial q_b}{\partial \tilde{q}_a} = -\frac{\partial U}{\partial \tilde{q}_a} \tag{2.3.35}$$

and the transformed equations of motion are

$$\frac{\mathrm{d}}{\mathrm{d}t}\left(\frac{\partial L}{\partial \tilde{v}_a}\right) - \frac{\partial L}{\partial \tilde{q}_a} = 0, \qquad \frac{\mathrm{d}}{\mathrm{d}t}(\tilde{q}_a) = \tilde{v}_a \tag{2.3.36}$$

where $L = T - U$. These are *Lagrange's equations.* An equivalent form is

$$\frac{\partial^2 L}{\partial \tilde{v}_a \partial \tilde{v}_b}\ddot{\tilde{q}}_b + \frac{\partial^2 L}{\partial \tilde{v}_a \partial \tilde{q}_b}\dot{\tilde{q}}_b + \frac{\partial^2 L}{\partial \tilde{v}_a \partial \tilde{t}} - \frac{\partial L}{\partial \tilde{q}_a} = 0, \qquad \dot{\tilde{q}}_a = \tilde{v}_a. \tag{2.3.36$'$}$$

Definition (2.3.3). The function $L = T - U$ on PT is the *Lagrangian.*

It is important to be able to recognize conservative forces. The first point is that if they are to be conservative, then the F_a can depend only on $q_1, \ldots, q_n$ and t, but not, for example, on $v_1, \ldots, v_n$: given the configuration and the time, but not the state of motion, it must be possible to determine the forces (there are generalizations for velocity dependent forces, but we shall not consider them).

The second point is concerned with the origin of the term 'conservative'. Consider, as before, imaginary displacements of the system at a fixed time t. Suppose that we move the particles through a continuous sequence of configurations represented by a curve $q_a = q_a(s)$ in C, where $s \in [s_1, s_2]$ is some parameter (Fig. 2.3.5). Then $\boldsymbol{r}_\alpha = \boldsymbol{r}_\alpha(s)$ and the total work done by the forces during this displacement is

$$W = \sum_\alpha \int_{s_1}^{s_2} \boldsymbol{F}_\alpha \cdot \frac{\mathrm{d}\boldsymbol{r}_\alpha}{\mathrm{d}s}\,\mathrm{d}s = \int_{s_1}^{s_2} F_a(q(s), t)\frac{\mathrm{d}q_a}{\mathrm{d}s}\,\mathrm{d}s. \tag{2.3.37}$$

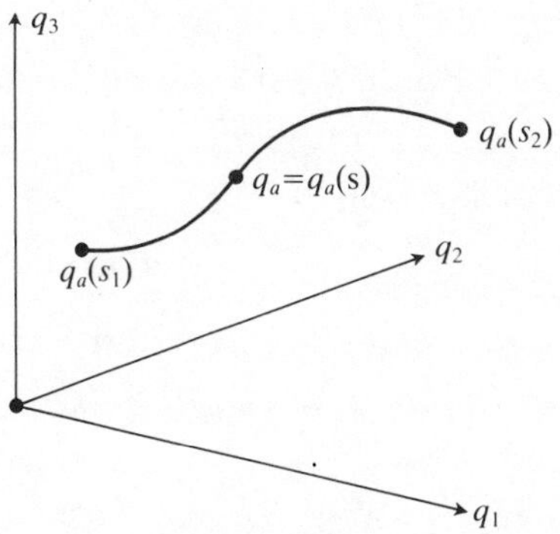

Fig. 2.3.5

If the forces are conservative with potential $U(q, t)$, then

$$W = -\int_{s_1}^{s_2} \frac{\partial U}{\partial q_a}\frac{\mathrm{d}q_a}{\mathrm{d}s}\,\mathrm{d}s = -\int_{s_1}^{s_2} \frac{\mathrm{d}}{\mathrm{d}s}(U(q(s), t))\,\mathrm{d}s$$
$$= U(q(s_1), t) - U(q(s_2), t), \tag{2.3.38}$$

and so the work done depends only on $q_a(s_1)$ and $q_a(s_2)$: it depends only on the initial and final configurations and not on the intermediate configurations.

The converse is also true: if for any curve in C, the work done in displacing the system along the curve (at fixed t) depends only on the endpoints of the curve, then the forces are conservative. Equivalently, the forces are conservative if the work done in displacing the system around any closed curve vanishes (a curve is closed if its two endpoints coincide). Hence the term 'conservative': for such curves, there is no net expenditure of energy.

One can see from this that gravitational and elastic forces are conservative: for example, no net work is done by gravity in lifting a particle and then returning it to its initial position; and no net work is done by elastic forces in stretching and then relaxing a perfectly elastic string. However, friction is not conservative.

Equation (2.3.38) also gives an interpretation of U: the difference in the values of U at two different configurations (at some time t) is the energy required to change the first configuration into the second. Thus in problems involving gravitational or elastic forces, U is the potential energy (plus an arbitrary constant).

There exists a potential $U(q, t)$ for forces $F_a(q, t)$ if and only if

$$\frac{\partial F_a}{\partial q_b} = \frac{\partial F_b}{\partial q_a}. \tag{2.3.39}$$

This *integrability condition* is sometimes useful.

Example (2.3.4). If the z-axis is vertical and the force $\boldsymbol{F}$ in example (2.3.3) is gravity, then $U = mgz = mgr\cos\theta$. From this, we can read off

$$\begin{aligned} &F_1 = F_2 = 0, \qquad F_3 = -mg \\ &\tilde{F}_1 = -\frac{\partial U}{\partial r} = -mg\cos\theta, \qquad \tilde{F}_2 = -\frac{\partial U}{\partial \theta} = mgr\sin\theta \\ &\tilde{F}_3 = -\frac{\partial U}{\partial \varphi} = 0. \end{aligned} \tag{2.3.40}$$

□

Example (2.3.5). For a single particle, put

$$\tilde{q}_1 = q_1 - tu, \qquad \tilde{q}_2 = q_2, \qquad \tilde{q}_3 = q_3, \qquad \tilde{t} = t. \tag{2.3.41}$$

Then the $\tilde{q}_a$ are Cartesian coordinates of a new inertial frame $\tilde{R}$, which is moving relative to the original frame R at constant speed u in the q_1-direction.

In the new coordinates, $\tilde{v}_1 = v_1 - u$ and

$$T = \tfrac{1}{2}m(\tilde{v}_1^2 + \tilde{v}_2^2 + \tilde{v}_3^2 + 2u\tilde{v}_1 + u^2). \tag{2.3.42}$$

This is slightly alarming: if we had begun by referring the motion to $\tilde{R}$ instead of R, then we should have started by writing

$$\tilde{T} = \tfrac{1}{2}m(\tilde{v}_1^2 + \tilde{v}_2^2 + \tilde{v}_3^2), \tag{2.3.43}$$

which is the kinetic energy relative to $\tilde{R}$. However, the discrepancy is not serious since

$$\frac{\mathrm{d}}{\mathrm{d}t}\left(\frac{\partial T}{\partial v_a}\right) = \frac{\mathrm{d}}{\mathrm{d}t}\left(\frac{\partial \tilde{T}}{\partial \tilde{v}_a}\right), \qquad \frac{\partial \tilde{T}}{\partial \tilde{q}_a} = \frac{\partial T}{\partial q_a} = 0, \tag{2.3.44}$$

so the equations of motion are the same, whichever route is followed. □

Example (2.3.6). We can transform from the Cartesian coordinates q_1, q_2, q_3 of an inertial frame $R = (O, B)$ to the Cartesian coordinates $\tilde{q}_1$, $\tilde{q}_2$, $\tilde{q}_3$ of a uniformly rotating frame $\tilde{R} = (O, \tilde{B})$ by writing

$$\begin{aligned} \tilde{q}_1 &= q_1\cos\omega t + q_2\sin\omega t \\ \tilde{q}_2 &= -q_1\sin\omega t + q_2\cos\omega t \\ \tilde{q}_3 &= q_3 \end{aligned} \tag{2.3.45}$$

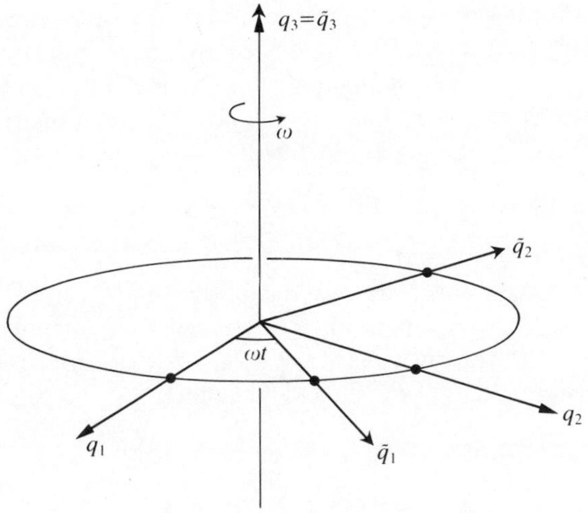

Fig. 2.3.6

where ω is constant. The $\tilde{q}$-axes are rotating relative to the q-axes with angular velocity ω about the q_3-axis (Fig. 2.3.6).

The corresponding velocity coordinates are related by

$$\begin{aligned}\tilde{v}_1 &= v_1 \cos \omega t + v_2 \sin \omega t - \omega q_1 \sin \omega t + \omega q_2 \cos \omega t,\\ \tilde{v}_2 &= -v_1 \sin \omega t + v_2 \cos \omega t - \omega q_1 \cos \omega t - \omega q_2 \sin \omega t,\\ \tilde{v}_3 &= v_3.\end{aligned} \tag{2.3.46}$$

For a single particle, the kinetic energy relative to R is

$$\begin{aligned}T &= \tfrac{1}{2}m(v_1^2 + v_2^2 + v_3^2)\\ &= \tfrac{1}{2}m(\tilde{v}_1^2 + \tilde{v}_2^2 + \tilde{v}_3^2 - 2\omega\tilde{q}_2\tilde{v}_1 + 2\omega\tilde{q}_1\tilde{v}_2 + \omega^2\tilde{q}_1^2 + \omega^2\tilde{q}_2^2).\end{aligned} \tag{2.3.47}$$

Hence the equations of motion in the absence of forces are

$$\begin{aligned} m\frac{\mathrm{d}}{\mathrm{d}t}(\tilde{v}_1 - \omega\tilde{q}_2) - m\omega\tilde{v}_2 - m\omega^2\tilde{q}_1 = 0 \qquad & \frac{\mathrm{d}\tilde{q}_1}{\mathrm{d}t} = \tilde{v}_1\\ m\frac{\mathrm{d}}{\mathrm{d}t}(\tilde{v}_2 + \omega\tilde{q}_1) + m\omega\tilde{v}_1 - m\omega^2\tilde{q}_2 = 0 \qquad & \frac{\mathrm{d}\tilde{q}_2}{\mathrm{d}t} = \tilde{v}_2\\ m\frac{\mathrm{d}}{\mathrm{d}t}(\tilde{v}_3) = 0 \qquad & \frac{\mathrm{d}\tilde{q}_3}{\mathrm{d}t} = \tilde{v}_3.\end{aligned} \tag{2.3.48}$$

It is easy to see that these are equivalent to

$$m\ddot{\boldsymbol{r}} + 2m\boldsymbol{\omega} \wedge \dot{\boldsymbol{r}} + m\boldsymbol{\omega} \wedge (\boldsymbol{\omega} \wedge \boldsymbol{r}) = 0 \tag{2.3.49}$$

where $\boldsymbol{r}$ is the position vector from O, $\boldsymbol{\omega} = \omega \boldsymbol{e}_3$, and the dot denotes the time derivative with respect to the non-inertial frame. □

Now that they have served their purpose, we shall drop the tildes: henceforth the q_a's and v_a's will be arbitrary generalized coordinates and velocities. At times we shall also use a common, but potentially dangerous, shorthand notation, in which v_a is replaced by $\dot{q}_a$, so that, for example, Lagrange's equations become

$$\frac{\mathrm{d}}{\mathrm{d}t}\left(\frac{\partial L}{\partial \dot{q}_a}\right) - \frac{\partial L}{\partial q_a} = 0. \tag{2.3.50}$$

The danger is that of confusing two meanings of '$\dot{q}_a$': at one stage it is a generalized velocity coordinate on PT, treated as independent of q_a and t; at another, it is the time derivative $\mathrm{d}q_a/\mathrm{d}t$. The q_a, v_a notation, in which the two meanings are clearly distinguished, is safer, but more cumbersome.

Exercises

(2.3.1) A particle of mass m is subject to a force $\boldsymbol{F}$. Obtain the equations of motion in cylindrical polar coordinates.

(2.3.2) A particle of unit mass is subject to an inverse-square-law central force

$$\boldsymbol{F} = -\frac{\boldsymbol{r}}{r^3}$$

where $r = |\boldsymbol{r}|$ and $\boldsymbol{r}$ is the position vector from the origin of an inertial frame. Show that the motion is governed by the Lagrangian

$$L = \tfrac{1}{2}\dot{\boldsymbol{r}} \cdot \dot{\boldsymbol{r}} + \frac{1}{r}.$$

Write down the equations of motion in spherical polar coordinates and show that there are solutions with $\theta = \pi/2$ throughout the motion.

(2.3.3) Two particles, each of mass m, are moving under their mutual gravitational attraction, which is given by the potential $U = -\gamma m/2r$, where $2r$ is their separation and γ is a constant. Find the equations of motion in terms of the coordinates X, Y, Z, r, θ, φ, where X, Y, and Z are the Cartesian coordinates of the centre of mass and r, θ, and φ are the polar coordinates of one particle relative to the centre of mass.

(2.3.4) The dynamics of a system with n degrees of freedom are governed by a Lagrangian $L(q, v, t)$. Show that if $f(q, t)$ is any function on CT, then

$$L' = L + \frac{\partial f}{\partial q_a} v_a + \frac{\partial f}{\partial t}$$

generates the same dynamics.

(2.3.5) A system has Lagrangian $L=\frac{1}{2}T_{ab}v_a v_b$ where the T_{ab} are functions of the q_a alone. Show that

$$\frac{dL}{dt}=0.$$

Show that if f is any function of one variable, then $L'=f(L)$ generates the same dynamics.

2.4 Holonomic constraints

In a wide class of mechanical problems we know that the motion of the system satisfies certain constraints, but we are not interested in the forces that maintain the constraints: all that we want to understand is the behaviour of the remaining unconstrained degrees of freedom. The Lagrangian formalism is particularly useful in this type of problem in that it provides a simple general method for eliminating unknown and unwanted constraint forces.

Before turning to the details, let us consider some examples.

Example (2.4.1). A particle confined to the surface of a smooth sphere of radius a. Here there is one constraint equation

$$x^2+y^2+z^2=a^2 \tag{2.4.1}$$

and two remaining degrees of freedom (corresponding to the two polar angles θ and φ). The force that maintains the constraint is the normal reaction $\boldsymbol{N}$. To find the motion of the particle, we need two differential equations for θ and φ which do not involve $\boldsymbol{N}$. □

Example (2.4.2). A bead sliding along a smooth rotating wire. Here there are two constraint equations and one degree of freedom (the position of the bead on the wire).

Example (2.4.3). A rigid body. In an idealized, and somewhat unrealistic model, a rigid body is a large collection of N particles with position vectors $\boldsymbol{r}_\alpha$ subject to the constraints

$$|\boldsymbol{r}_\alpha-\boldsymbol{r}_\beta|=\text{constant}. \tag{2.4.2}$$

These reduce the number of degrees of freedom from $3N$ to six (three for the position of the centre of mass and three for rotations about the centre of mass).

Example (2.4.4). A rigid sphere rolling on a rough plane. Here, in addition to eqn (2.4.2), we have the extra condition

$$\dot{\boldsymbol{r}} + \boldsymbol{\omega} \wedge (-a\boldsymbol{k}) = 0 \tag{2.4.3}$$

where $\boldsymbol{r}$ is the position vector of the centre, $\boldsymbol{\omega}$ is the angular velocity of the sphere, a is the radius, and $\boldsymbol{k}$ is the unit vector normal to the plane. This is the condition that the particle of the sphere in contact with the plane should be instantaneously at rest. □

In the first and third examples, the constraint equations are of the form

$$f(q) = 0. \tag{2.4.4}$$

In the second example, we have a *moving* constraint: the condition that the bead remains on the wire is given by two equations of the form

$$f(q, t) = 0 \tag{2.4.5}$$

where the q_a are the Cartesian coordinates of the bead. In the final example, the constraint equations also involve the velocities of the particles making up the sphere: eqn (2.4.3) gives rise to two additional equations of the form

$$f(q, v, t) = 0. \tag{2.4.6}$$

Constraints like eqns (2.4.4) and (2.4.5) that do not involve the velocities are said to be *holonomic*. Holonomic constraints are further subdivided into *fixed* or *scleronomic* constraints that do not involve t (as in eqn (2.4.4)) and *moving* or *rheonomic* constraints that do involve t (as in eqn (2.4.5)). For the moment we shall deal only with the holonomic constraints.

Suppose that we have a system of N particles (as in sections 2.2 and 2.3) subject to $n - m$ constraints of the form

$$f_r(q, t) = 0; \qquad r = 1, 2, \ldots, n - m \tag{2.4.7}$$

where the q_a are generalized coordinates. We shall think of these as defining an $(m + 1)$-dimensional *constraint surface* in CT (Fig. 2.4.1).

Provided that the constraints are independent, we should be able to use eqns (2.4.7) to determine the configuration of the system from a knowledge of t and m of the coordinates q_a: the constraints should reduce the number of degrees of freedom from $n = 3N$ to m. This is made precise by the following.

Definition (2.4.1). The constraints $f_r = 0$ are *independent* if the $(n - m) \times n$ matrix with entries $\partial f_r / \partial q_a$ has maximal rank at each point.

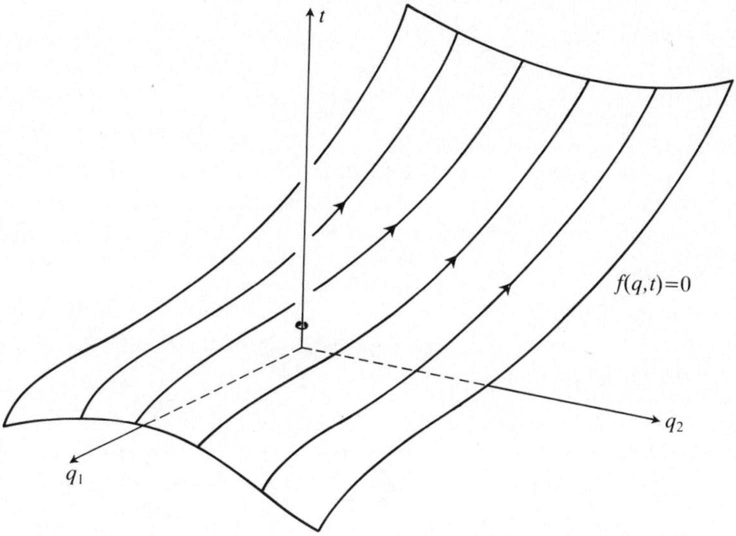

Fig. 2.4.1 The constraint surface ($n = 2$, $m = 1$).

Proposition (2.4.1). If the constraints $f_r = 0$ are independent, then there exists a system of generalized coordinates $\tilde{q}_a$ in which the constraints are

$$\tilde{q}_{m+1} = 0, \qquad \tilde{q}_{m+2} = 0, \ldots, \tilde{q}_n = 0. \tag{2.4.8}$$

Independence precludes meaningless constraints (such as $t = 0$) and prevents double counting (it fails, for example, if $f_r = f_s$ for some $s \neq r$). We shall not go through the proof of the proposition (it would be a simple exercise in any course on advanced calculus): all that is needed is to find m functions on CT which, together with t and the f_r's themselves, make up a coordinate system on CT.

Definition (2.4.2). If eqn (2.4.8) holds, then the generalized coordinates $\tilde{q}_a$ are said to be *adapted* to the constraints.

Example (2.4.5). For the particle moving on the surface of a sphere, the coordinates $q_1 = \theta$, $q_2 = \varphi$, $q_3 = r - a$ are adapted to the constraint $x^2 + y^2 + z^2 - a^2 = 0$. □

Suppose that we can write the force $\boldsymbol{F}_\alpha$ on each particle as the sum of two terms

$$\boldsymbol{F}_\alpha = \boldsymbol{E}_\alpha + \boldsymbol{K}_\alpha \tag{2.4.9}$$

where the $\boldsymbol{E}_\alpha$ are known external forces (such as gravity) and the $\boldsymbol{K}_\alpha$ are *constraint forces* responsible for maintaining the constraints (in the first example (2.4.1), the normal reaction of the sphere on the particle is a constraint force; as are the normal reaction of the wire on the bead in the second example, and the forces between the particles in the rigid body in the third example).

Then we can also write

$$F_a = E_a + K_a \tag{2.4.10}$$

where the E_a are the q-components of the $\boldsymbol{E}_\alpha$ and the K_a are the q-components of the $\boldsymbol{K}_\alpha$.

The only assumption that we shall make about the constraint forces $\boldsymbol{K}_\alpha$ is that they are *workless* in the sense that *they do no work during any instantaneous displacement of the system consistent with the constraints.* The precise meaning of this assumption must be clearly understood. Fix a time t, and imagine that at this time, we take a photograph of the system with a camera that records not only the positions of the individual particles, but also the forces acting on them (as in Fig. 2.4.2). Now imagine making a small change in the recorded configuration in which $q_a \mapsto q_a + \delta q_a$ with t held fixed; in other words, in which particle α is moved from the point with position vector $\boldsymbol{r}_\alpha$ to the point with position vector $\boldsymbol{r}_\alpha + \delta \boldsymbol{r}_\alpha$ where

$$\delta \boldsymbol{r}_\alpha = \frac{\partial \boldsymbol{r}_\alpha}{\partial q_a} \delta q_a \tag{2.4.11}$$

(ignoring second-order terms in δq_a). The work done by the constraint

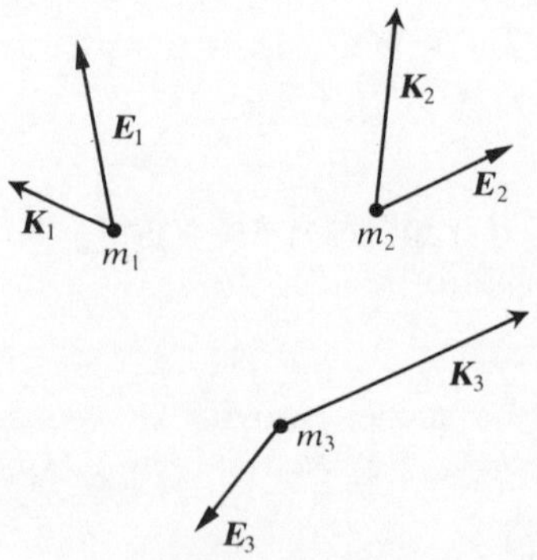

Fig. 2.4.2

forces is

$$W = \sum_{\alpha} \boldsymbol{K}_\alpha \cdot \delta \boldsymbol{r}_\alpha = K_a \delta q_a \tag{2.4.12}$$

(to the first order).

The assumption is that $W = 0$ whenever the displacement is consistent with the constraints. This means the displaced configuration must also be one that satisfies the constraint equations at time t; that is,

$$0 = f_r(q + \delta q, t) = \delta q_a \frac{\partial f_r}{\partial q_a} \tag{2.4.13}$$

to the first order (since $f_r(q, t) = 0$). Thus our assumption is that

$$K_a \delta q_a = 0 \quad \text{whenever} \quad \delta q_a \frac{\partial f_r}{\partial q_a} = 0 \tag{2.4.14}$$

to the first order in the δq_a.

We can express this in a less clumsy way: put $\delta q_a = \varepsilon X_a$, where $X_a \in \mathbb{R}$ and ε is a small parameter. Then, by cancelling ε, eqn (2.4.14) becomes

$$K_a X_a = 0 \quad \text{whenever} \quad X_a \frac{\partial f_r}{\partial q_a} = 0, \tag{2.4.15}$$

which is now required to hold exactly, without the qualification 'to the first order'.

If the generalized coordinates q_a are adapted to the constraints, then the second of eqns (2.4.14) is equivalent to

$$\delta q_{m+1} = \delta q_{m+2} = \cdots = \delta q_n = 0. \tag{2.4.16}$$

In this case the condition that the constraint forces should be workless reduces to

$$K_1 = K_2 = \cdots = K_m = 0$$

and the equations of motion for $a = 1, 2, \ldots, m$ become

$$\frac{\mathrm{d}}{\mathrm{d}t}\left(\frac{\partial T}{\partial v_a}\right) - \frac{\partial T}{\partial q_a} = E_a, \qquad \frac{\mathrm{d}}{\mathrm{d}t} q_a = v_a. \tag{2.4.17}$$

These combine to give m second-order equations (the number needed) in which the constraint forces do not appear.

It is sometimes convenient to work in coordinates that are not adapted to the constraints. We can then still write

$$X_a\left[\frac{\mathrm{d}}{\mathrm{d}t}\left(\frac{\partial T}{\partial v_a}\right) - \frac{\partial T}{\partial q_a} - E_a\right] = 0 \quad \text{whenever} \quad X_a \frac{\partial f_r}{\partial q_a} = 0. \tag{2.4.18}$$

This is a form of *d'Alembert's principle.*

It is important to understand two points. First, in the case of moving constraints, when the f_r depend on time, a displacement satisfying eqn

(2.4.13) need not represent a possible motion of the system (whatever the external forces): if in time δt, the configuration changes from q_a to $q_a + \delta q_a$ during the actual motion, then the later configuration will satisfy $f_r(q + \delta q, t + \delta t) = 0$; that is

$$0 = \frac{\partial f_r}{\partial q_a} \delta q_a + \frac{\partial f_r}{\partial t} \delta t \qquad (2.4.19)$$

to the first order in δq_a and δt. In general this is incompatible with eqn (2.4.13).

Second, one can give no general justification for the assumption: it is simply that in a large number of examples, and, in particular in the first three cited above, one can split the forces $\boldsymbol{F}_\alpha$ into a sum of known external forces $\boldsymbol{E}_\alpha$ and unknown constraint forces $\boldsymbol{K}_\alpha$ that can easily be seen to satisfy eqn (2.4.14).

Example (2.4.6). For the bead sliding on a rotating wire, the constraint force is the normal reaction $\boldsymbol{N}$. If we freeze the motion of the wire and slide the bead a small distance along the wire, then $\boldsymbol{N}$ does no work since it is orthogonal to the wire. Hence the moving constraints in this system are workless. This is in spite of the fact that $\boldsymbol{N}$ *does* do work during the actual motion, since the rotation contributes to the velocity of the bead a component normal to the wire. □

It is not actually necessary to find *all* the coordinates q_a to make use of eqn (2.4.17); and, for example, in the case of a rigid body made up of 10^{25} particles, it would not be practical in anyway. With the coordinates adapted to the constraints, we know that $q_a = v_a = 0$ for $a = m+1, \ldots, n$ throughout the actual motion. To write down eqn (2.4.17), therefore, we need only know the expression for T when $q_a = v_a = 0$ for $a = m+1, \ldots, n$. So we can obtain the equations of motion for the remaining coordinates $q_1, \ldots, q_m$ by doing the following.

(1) Choose coordinates $q_1, q_2, \ldots, q_m$ that label the configurations satisfying the constraints.
(2) Express T in terms of $q_1, \ldots, q_m, v_1, \ldots, v_m$, and t on the assumption that the constraints are satisfied at all times.
(3) Find the q-components of the external forces by considering, for example, a small displacement consistent with the constraints of the form

$$q_1 \mapsto q_1 + \delta q_1, \qquad q_2 \mapsto q_2, \ldots, q_m \mapsto q_m \qquad (2.4.20)$$

with t held fixed, and equating $E_1 \delta q_1$ to the work done by the external forces.

When the external forces are conservative, the third step is much simpler: in this case there is a potential U for the $\boldsymbol{E}_\alpha$ and we can write

$$E_a = -\frac{\partial U}{\partial q_a} \qquad a = 1, 2, \ldots m. \tag{2.4.21}$$

The first of eqns (2.4.17) then reduces to the Lagrangian form

$$\frac{\mathrm{d}}{\mathrm{d}t}\left(\frac{\partial L}{\partial v_a}\right) - \frac{\partial L}{\partial q_a} = 0 \tag{2.4.22}$$

where $L = T - U$ and a runs over $1, 2, \ldots, m$.

Definition (2.4.3). The integer m is the number of *residual degrees of freedom.*

Unless otherwise stated, the q_a in problems with constraints will always be adapted coordinates; and the Einstein conventions will apply for $a, b, \ldots = 1, 2, \ldots, m$, where m is the number of residual degrees of freedom.

The q_a for $a = 1, 2, \ldots, m$ are coordinates in the *constrained configuration space*; q_a and v_a $(a = 1, \ldots, m)$ are coordinates in the *constrained phase space,* and so on. When there is no danger of confusion, the qualifications 'constrained' and 'residual' will often be dropped; and the notation C, CT, P, and PT will be used for the constrained spaces associated with the residual degrees of freedom.

Example (2.4.7). Consider a particle of mass m moving on a smooth sphere of radius a, with gravity as the only external force. The constraint force is the normal reaction, which is workless (since it does no work during small displacements of the particle along the surface).

Here we can take (Fig. 2.4.3)

$$q_1 = \theta, \qquad q_2 = \varphi, \tag{2.4.23}$$

in terms of which the kinetic energy and the gravitational potential are given by

$$\begin{aligned} T &= \tfrac{1}{2}ma^2(\dot\theta^2 + \dot\varphi^2 \sin^2\theta) \\ U &= mga \cos\theta. \end{aligned} \tag{2.4.24}$$

Thus the equations of motion are

$$\begin{aligned} \frac{\mathrm{d}}{\mathrm{d}t}(ma^2\dot\theta) - ma^2\dot\varphi^2 \cos\theta \sin\theta - mga \sin\theta = 0 \\ \frac{\mathrm{d}}{\mathrm{d}t}(ma^2\dot\varphi \sin^2\theta) = 0. \end{aligned} \tag{2.4.25}$$

Example (2.4.8). In example (1.3.1), the bead has one (residual) degree of freedom corresponding to the coordinate $q = \theta$. The velocity of the

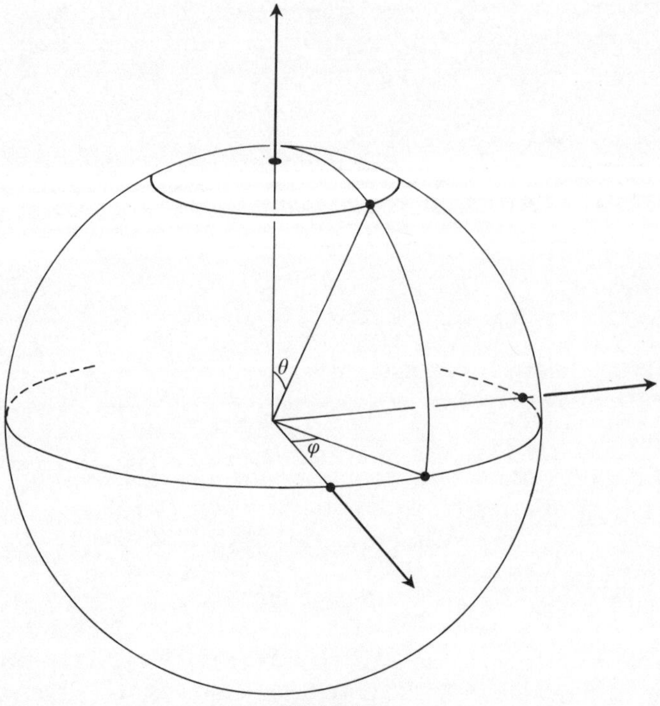

Fig. 2.4.3

bead relative to the inertial frame is

$$\dot{\boldsymbol{r}} + \boldsymbol{\omega} \wedge \boldsymbol{r} = a\dot{\theta}(-\sin\theta\, \boldsymbol{i} + \cos\theta\, \boldsymbol{j}) \\ + \omega a(\sin\alpha \sin\theta\, \boldsymbol{k} + \cos\alpha\cos\theta\, \boldsymbol{j} - \cos\alpha\sin\theta\, \boldsymbol{i}). \tag{2.4.26}$$

Hence

$$T = \tfrac{1}{2}ma^2\dot{\theta}^2 + ma^2\omega\dot{\theta}\cos\alpha + \tfrac{1}{2}ma^2\omega^2(\sin^2\alpha\sin^2\theta + \cos^2\alpha) \tag{2.4.27}$$

and

$$U = mga\sin\alpha\cos\theta. \tag{2.4.28}$$

Therefore the equation of motion is

$$ma^2\frac{\mathrm{d}}{\mathrm{d}t}(\dot{\theta} + \omega\cos\alpha) - ma^2\omega^2\sin^2\alpha\sin\theta\cos\theta - mga\sin\alpha\sin\theta = 0 \tag{2.4.29}$$

in agreement with the result derived by vector methods. □

The constrained configuration and phase spaces can have complicated topological properties and it may not be possible to use only a single coordinate system. For example, for the particle moving on a sphere, the polar coordinates θ and φ are singular on the polar axis at $\theta = 0$. Other coordinate systems must be used to describe motion near the north and south poles.

Example (2.4.9). A simple pendulum has the equation of motion

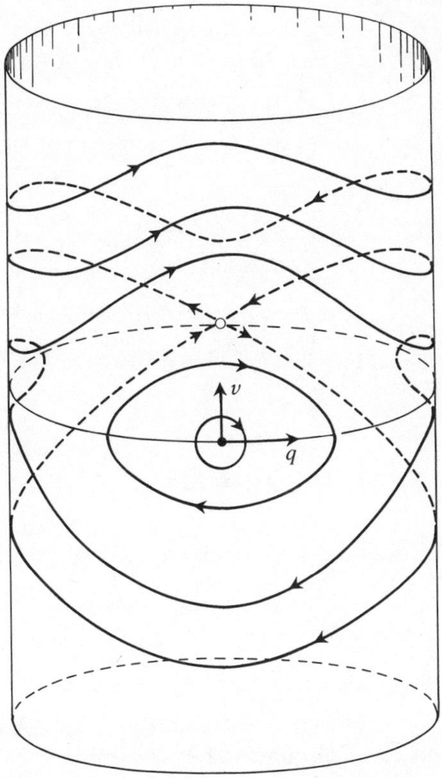

Fig. 2.4.4 The phase space of the simple pendulum. The point • is the stable equilibrium; the point o (on the far side of the cylinder) is the unstable equilibrium in which the pendulum is at rest in the uppermost position ($q = \pi$).

$\ddot{\theta} = -k \sin\theta$, where θ is the angle with the vertical: we can think of the system as consisting of a single particle constrained to move on a circle in a vertical plane.

The (constrained) configuration space is a circle and the (constrained) phase space is a cylinder, on which $q = \theta$ increases around the circular cross-sections and $v = \dot{\theta}$ increases along the generators. The orbits in the phase space are illustrated in Fig. 2.4.4.

Exercises

(2.4.1) A particle of mass m is free to move in a horizontal plane. It is attached to a fixed point O by a light elastic string, of natural length a and modulus λ. Show that the tension in the string is conservative and show that the Lagrangian for the motion when $r > a$ is

$$L = \tfrac{1}{2}m(\dot{r}^2 + r^2\dot{\theta}^2) - \frac{\lambda}{2a}(r-a)^2$$

where r and θ are plane polar coordinates with origin O.

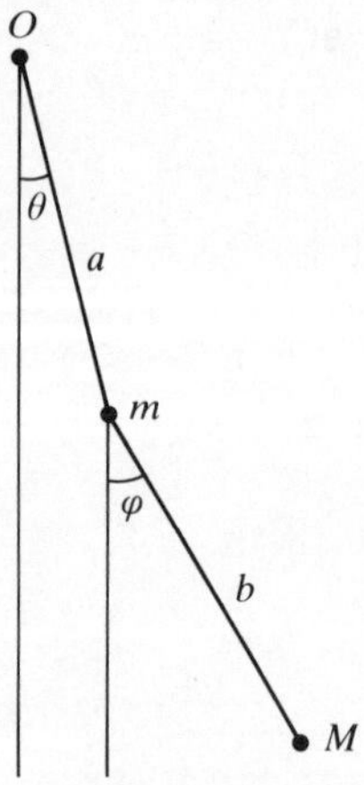

Fig. 2.4.5

(2.4.2) The double pendulum (Fig. 2.4.5) is free to move in a vertical plane. Show that the motion is governed by the Lagrangian

$$L = \tfrac{1}{2}ma^2\dot{\theta}^2 + \tfrac{1}{2}M(a^2\dot{\theta}^2 + b^2\dot{\varphi}^2 + 2ab\dot{\theta}\dot{\varphi}\cos(\theta - \varphi))$$
$$+ mga\cos\theta + Mg(b\cos\varphi + a\cos\theta).$$

(2.4.3) A particle of mass m is constrained to move under gravity on the surface of a smooth right circular cone of semi-vertical angle $\pi/4$. The axis of the cone is vertical, with the vertex downwards. Find the equations of motion in terms of z (the height above the vertex) and θ (the angular coordinate around the circular cross-sections). Show that

$$\dot{z}^2 + \frac{h^2}{2z^2} + gz = E$$

where E and h are constant. Sketch and interpret the orbits in the z, $\dot{z}$-plane for a fixed value of h.

(2.4.4) A particle P of mass m is attached to two light inextensible strings, each of length a. The strings pass over two smooth pegs A and B, which are at the same height and distance $2b$ apart. At the other ends of the strings hang two particles of mass m, which can move up and down the vertical lines through A and B. The particle P can move in the vertical plane containing A and B.

Show that if $2b\cosh\varphi = PA + PB$ and $2b\cos\theta = PA - PB$, then the kinetic energy of P is

$$T = \tfrac{1}{2}mb^2(\cosh^2\varphi - \cos^2\theta)(\dot{\theta}^2 + \dot{\varphi}^2).$$

Hence find the Lagrangian of the system in terms of θ and φ.

2.5* Symmetry and momentum

In spherical polar coordinates, the motion of a free particle is governed by the Lagrangian

$$L = \tfrac{1}{2}m(\dot{r}^2 + r^2\dot{\theta}^2 + r^2\dot{\varphi}^2 \sin^2\theta). \tag{2.5.1}$$

Since $\partial L/\partial \varphi = 0$, we can read off from Lagrange's equations that the angular momentum about the z-axis,

$$mr^2\dot{\varphi}\sin^2\theta = \frac{\partial L}{\partial \dot{\varphi}}, \tag{2.5.2}$$

is conserved.

As a generalization, suppose that we have a system of particles subject to conservative forces and that we have expressed the Lagrangian $L = T - U$ as a function of t and of the generalized position and velocity coordinates q_a and v_a ($a = 1, 2, \ldots, n$). If L is independent of q_a for some a, then

$$p_a = \frac{\partial L}{\partial v_a} \tag{2.5.3}$$

is constant along the orbits in PT.

> **Definition (2.5.1).** The quantity p_a is the *generalized momentum* conjugate to q_a. A coordinate q_a such that $\partial L/\partial q_a = 0$ is said to be *cyclic*.

Once again the second fundamental confusion raises its head. For example, $\partial L/\partial q_1 = 0$ is a property not just of the coordinate q_1, but of the entire coordinate system $q_1, \ldots, q_n$; it is possible to introduce new coordinates $\tilde{q}_a$ such that $\tilde{q}_1 = q_1$, but $\partial L/\partial \tilde{q}_1 \neq \partial L/\partial q_1$. It can also happen that $\partial L/\partial \tilde{v}_1 \neq \partial L/\partial v_1$, so that the momentum conjugate to $\tilde{q}_1 = q_1$ need not be the same in the two coordinate systems.

For small ε, the quantity εp_1 is the energy needed to change the generalized velocity v_1 to $v_1 + \varepsilon$, keeping the configuration and the other generalized velocities fixed. This is the sense in which p_1 is a 'momentum' (linear momentum is a measure of how much energy is needed to change linear velocity by a small amount; angular momentum is a measure of how much energy is required to alter angular velocity).

A cyclic coordinate corresponds to a symmetry in the dynamics of the

* The only parts of this section that will be used later in the book are: the definition of cyclic coordinates; proposition (2.5.2); and the principles of linear and angular momentum.

system. The conservation of $mr^2\dot{\varphi}\sin^2\theta$ in our example arises from the fact that the Lagrangian and the equations of motion are unchanged by adding a constant to φ; that is, from the invariance of the dynamics under rotation about the z-axis.

But there is nothing special about the z-axis. We could just as well measure the spherical polar coordinates from the x-axis. Then the cyclic coordinate would be the one associated with symmetry under rotations about the x-axis; and the corresponding conserved quantity would be angular momentum about the x-axis.

It would be useful to be able to obtain the conservation of angular momentum about the x-axis without actually having to make the transformation to a new system of spherical polar coordinates; and, in general, to be able to spot the presence of a cyclic coordinate of one coordinate system when the Lagrangian is written in terms of another set of generalized coordinates and velocities.

We can do this by representing a symmetry as a *vector field*. The general idea is illustrated by our example of a free particle. Under a rotation through a small angle ε about the z-axis, the position vector of the particle changes from $\boldsymbol{r}$ to $\boldsymbol{r}+\varepsilon\boldsymbol{X}$, where $\boldsymbol{X}=\boldsymbol{k}\wedge\boldsymbol{r}$ ($\boldsymbol{k}$ is the unit vector along the z-axis); $\boldsymbol{X}$ is a vector field on $\mathbb{R}^3$—the configuration space of the particle. We shall think of $\boldsymbol{X}$ as the *generator* of small rotations about the z-axis.

Not all vector fields on $\mathbb{R}^3$ generate symmetries. To understand what is special about this particular $\boldsymbol{X}$, we have to consider the effect of the rotation not only on the particle's position, but also on its state of motion. If the particle is moving with velocity $\boldsymbol{v}$ at the point with position vector $\boldsymbol{r}$, and if we rotate its orbit through ε about the z-axis, then its velocity changes to $\boldsymbol{v}+\varepsilon\boldsymbol{W}$ where

$$\boldsymbol{W}=\boldsymbol{k}\wedge\boldsymbol{v}=\frac{\mathrm{d}\boldsymbol{X}}{\mathrm{d}t}. \tag{2.5.4}$$

Thus $\boldsymbol{X}$ also generates small displacements in the phase space P, given by $\delta\boldsymbol{r}=\varepsilon\boldsymbol{X}$, $\delta\boldsymbol{v}=\varepsilon\boldsymbol{W}$. The property that characterizes $\boldsymbol{X}$ as a symmetry is that the particle's Lagrangian $L=\frac{1}{2}m\boldsymbol{v}\cdot\boldsymbol{v}$ is unchanged to the first order in ε by this small change in $\boldsymbol{r}$ and $\boldsymbol{v}$.

In this particular example, the symmetry is represented by a single vector field $\boldsymbol{X}$ on the configuration space of the particle. In a general system with n degrees of freedom, we shall look at symmetries associated with time-dependent vector fields on C—of which the time-independent vector field $\boldsymbol{X}$ can be considered a special case.

We shall define a 'time-dependent vector field' X on C in terms of its *components*, which will be n functions $X_a(q,t)$ on CT; and we shall associate with X small displacements in the configuration of the system in which (q_a,t) is displaced to $(q_a+\delta q_a,t)$ where $\delta q_a=\varepsilon X_a(q,t)$ for some small ε, with t held fixed. For example, if q_1, q_2, q_3 are the Cartesian coordinates of the free particle, then X_1, X_2, X_3 are the three components of $\boldsymbol{k} \wedge \boldsymbol{r}$ along the coordinate axes.

The key part of the definition is the transformation law for the components under change of coordinates. If $q'_a = q'_a(q, t)$ is a new system of generalized coordinates, then

$$\delta q'_a = \frac{\partial q'_a}{\partial q_b} \delta q_b + \mathrm{O}(\varepsilon^2). \tag{2.5.5}$$

Thus if X is to be associated with the *same* small displacements irrespective of the choice of coordinate system, then its components in the new system must be

$$X'_a = \frac{\partial q'_a}{\partial q_b} X_b. \tag{2.5.6}$$

Hence we arrive at the following.

Definition (2.5.2). A *time-dependent vector field* X on C is a family of functions $X_a(q, t)$ on CT associated with each generalized coordinate system, with the transformation law

$$X'_a = \frac{\partial q'_a}{\partial q_b} X_b$$

under change of generalized coordinates. The X_a are the *components* of X.

A vector field can be pictured as a field of arrows on C at each t, with the arrow at q_a pointing to $q_a + \varepsilon X_a(q, t)$. As t increases, the pattern of arrows on C changes. Alternatively, X can be pictured as a single field of arrows on CT, with the arrow at (q_a, t) pointing to $(q_a + \varepsilon X_a(q, t), t)$. See Fig. 2.5.1.

We can also represent X by a family of ordinary vectors $\boldsymbol{X}_\alpha$, one for each particle of the system. Under the displacement associated with X, the particle at $\boldsymbol{r}_\alpha = \boldsymbol{r}_\alpha(q, t)$ is moved to $\boldsymbol{r}_\alpha + \varepsilon \boldsymbol{X}_\alpha$, where

$$\boldsymbol{X}_\alpha = \frac{\partial \boldsymbol{r}_\alpha}{\partial q_b} X_b. \tag{2.5.7}$$

The $\boldsymbol{X}_\alpha$ depend on the configuration of all the particles and the time t. That is, $\boldsymbol{X}_\alpha = \boldsymbol{X}_\alpha(q, t)$.

A vector field generates displacements in the state of motion of the system. If $q_a = x_a(t)$ is the equation of a curve in CT representing a

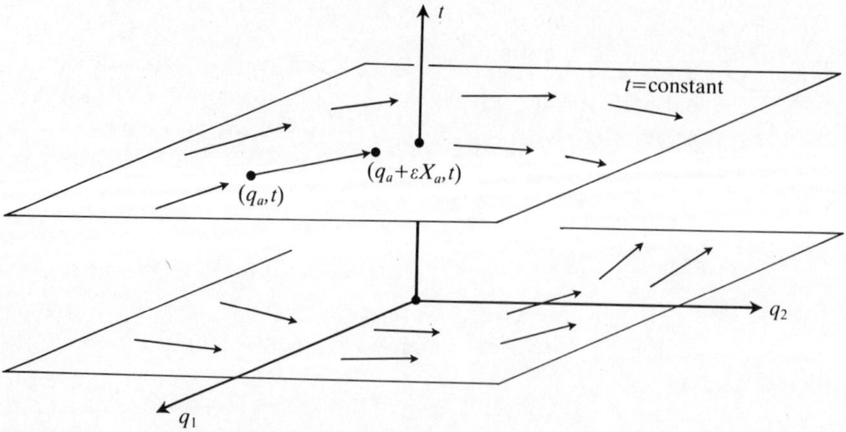

Fig. 2.5.1 A time-dependent vector field, pictured as a field of arrows on CT ($n=2$).

possible motion of the system, then

$$q_a = x_a(t) + \varepsilon X_a(x(t), t) \tag{2.5.8}$$

is the curve obtained by making the small displacements associated with X. Along its extension to PT,

$$v_a = \dot{x}_a(t) + \varepsilon W_a(x(t), \dot{x}(t), t) \tag{2.5.9}$$

where

$$W_a(q, v, t) = \frac{\mathrm{d}X_a}{\mathrm{d}t} = \frac{\partial X_a}{\partial q_b} v_b + \frac{\partial X_a}{\partial t} \tag{2.5.10}$$

Thus we associate with X the small displacements in PT under which (q_a, v_a, t) is sent to $(q_a + \varepsilon X_a, v_a + \varepsilon W_a, t)$.

By displacing a curve in CT by εX and then taking its extension to PT, we obtain the same result as by first taking the extension to PT and then making the displacement (to the first order in ε). Since the extension of a curve to PT is the same whatever coordinate system is used, the family of displacements in PT generated by X is also the same, irrespective of the choice of generalized coordinates.

In terms of the individual particles, the displacement in the state of motion is given by moving the particle at $\boldsymbol{r}_\alpha$ to $\boldsymbol{r}_\alpha + \varepsilon \boldsymbol{X}_\alpha$ and changing its velocity from $\boldsymbol{v}_\alpha$ to $\boldsymbol{v}_\alpha + \varepsilon \boldsymbol{W}_\alpha$, where

$$\boldsymbol{W}_\alpha = \frac{\mathrm{d}\boldsymbol{X}_\alpha}{\mathrm{d}t} = \frac{\partial \boldsymbol{X}_\alpha}{\partial q_b} v_b + \frac{\partial \boldsymbol{X}_\alpha}{\partial t}. \tag{2.5.11}$$

The vector field is a *dynamical symmetry* if the Lagrangian is invariant under this operation to the first order in ε; in other words if

$$L(q+\varepsilon X, v+\varepsilon W, t) = L(q, v, t) + O(\varepsilon^2). \tag{2.5.12}$$

By putting this in a slightly different way, we have the following.

Definition (2.5.3). The time-dependent vector field X on C is a *dynamical symmetry* if

$$X_a \frac{\partial L}{\partial q_a} + W_a \frac{\partial L}{\partial v_a} = 0 \tag{2.5.13}$$

where $W_a = v_b\, \partial X_a/\partial q_b + \partial X_a/\partial t$.

Definition (2.5.4). The *momentum conjugate* to X is the function $p: PT \to \mathbb{R}$

$$p = X_a \frac{\partial L}{\partial v_a}. \tag{2.5.14}$$

Proposition (2.5.1) Noether's theorem. If X is a dynamical symmetry of a system with Lagrangian $L(q, v, t)$, then its conjugate momentum is constant during the motion of the system.

Proof. By differentiating eqn (2.5.14) with respect to t along the orbits in PT,

$$\begin{aligned}\frac{\mathrm{d}p}{\mathrm{d}t} &= \frac{\mathrm{d}X_a}{\mathrm{d}t}\frac{\partial L}{\partial v_a} + X_a \frac{\mathrm{d}}{\mathrm{d}t}\left(\frac{\partial L}{\partial v_a}\right)\\ &= W_a \frac{\partial L}{\partial v_a} + X_a \frac{\partial L}{\partial q_a} \qquad (2.5.15)\\ &= 0.\end{aligned}$$

□

Example (2.5.1). Suppose that q_1 is cyclic. Define X by

$$X_1 = 1, \qquad X_2 = 0, \ldots, X_n = 0. \tag{2.5.16}$$

Then $W_a = 0$ and eqn (2.5.13) reduces to

$$\frac{\partial L}{\partial q_1} = 0. \tag{2.5.17}$$

Therefore X is a symmetry in the sense of the definition. The displacements generated by X are given by adding a constant ε to q_1. The conjugate momentum is $p_1 = \partial L/\partial v_1$.

Example (2.5.2). For a free particle in space: a small rotation about the x-axis is given by

$$\delta x = 0, \qquad \delta y = -\varepsilon z, \qquad \delta z = \varepsilon y. \tag{2.5.18}$$

If $q_1 = x$, $q_2 = y$, and $q_3 = z$, then the corresponding vector field X has components

$$X_1 = 0, \qquad X_2 = -q_3, \qquad X_3 = q_2. \tag{2.5.19}$$

The W_a are

$$W_1 = 0, \qquad W_2 = -v_3, \qquad W_3 = v_2. \tag{2.5.20}$$

This vector field is a dynamical symmetry of the Lagrangian

$$L = \tfrac{1}{2}m(v_1^2 + v_2^2 + v_3^2), \tag{2.5.21}$$

since eqn (2.5.13) reduces to

$$X_a \frac{\partial L}{\partial q_a} + W_a \frac{\partial L}{\partial v_a} = -v_3 \frac{\partial L}{\partial v_2} + v_2 \frac{\partial L}{\partial v_3} = 0. \tag{2.5.22}$$

The conjugate conserved momentum is

$$p = X_a \frac{\partial L}{\partial v_a} = -mz\dot{y} + my\dot{z}, \tag{2.5.33}$$

which is the x-component of the angular momentum $m\boldsymbol{r} \wedge \boldsymbol{v}$. □

It is natural to think of momentum as conjugate to a vector field rather than to a coordinate: the momentum conjugate to q_1 depends on the choice made for $q_2, \ldots, q_n$. But the momentum conjugate to X is independent of the choice of coordinates since

$$X'_a \frac{\partial L}{\partial v'_a} = X_a \frac{\partial q'_b}{\partial q_a} \frac{\partial L}{\partial v'_b} = X_a \frac{\partial L}{\partial v_a}. \tag{2.5.34}$$

Every dynamical symmetry X arises from a cyclic coordinate in some system of generalized coordinates. We shall not attempt to prove this, although it is a relatively simple exercise in advanced calculus.

A conserved momentum arises when L is invariant under small instantaneous displacements in the state of motion. Another conservation law arises when L is invariant under *time translation*: if L is such that $L(q, v, t + \varepsilon) = L(q, v, t)$, then there is a symmetry in the sense that the system behaves in the same way irrespective of the time at which it is set in motion.

We shall leave the full result until Chapter 4 and deal now only with a special case. Suppose that $L = T - U$, where $U = U(q, t)$ is the potential and T, the kinetic energy, is a homogeneous quadratic in the velocities:

$$T = \tfrac{1}{2}T_{ab}(q, t)v_a v_b. \tag{2.5.35}$$

Then $E = T + U$ is the *total energy* of the system.

Proposition (2.5.2). Let $L = T - U$ where T is a homogeneous quadratic in the velocities. If $\partial L/\partial t = 0$, then E is conserved.

Proof. Suppose that $\partial L/\partial t = 0$. Then $\partial U/\partial t = 0$ and $\partial T/\partial t = 0$. Lagrange's equations are

$$\frac{\partial^2 T}{\partial v_a\,\partial v_b}\frac{\mathrm{d}v_b}{\mathrm{d}t} + \frac{\partial^2 T}{\partial v_a\,\partial q_b}v_b - \frac{\partial T}{\partial q_a} + \frac{\partial U}{\partial q_a} = 0. \tag{2.5.36}$$

Hence, by multiplying by v_a and summing over a,

$$0 = \frac{\partial T}{\partial v_b}\frac{\mathrm{d}v_b}{\mathrm{d}t} + \frac{\partial T}{\partial q_b}v_b + v_a\frac{\partial U}{\partial q_a} = \frac{\mathrm{d}}{\mathrm{d}t}(T + U). \tag{2.5.37}$$

Here we have used

$$v_a\frac{\partial^2 T}{\partial v_a\,\partial v_b} = \frac{\partial T}{\partial v_b} \quad\text{and}\quad v_a\frac{\partial^2 T}{\partial v_a\,\partial q_b} = 2\frac{\partial T}{\partial q_b} \tag{2.5.38}$$

which follow from Euler's theorem on homogeneous functions: $\partial T/\partial v_b$ is homogeneous of degree one in the velocities; $\partial T/\partial q_b$ is homogeneous of degree two in the velocities. □

Noether's theorem shows that the momentum conjugate to a symmetry is a constant of the motion. There is a weaker form of symmetry which, while it does not give rise directly to constants of the motion, is still useful in the analysis of nonconservative systems.

Suppose that a system of particles has kinetic energy $T = T(q, v, t)$ (in generalized coordinates).

Definition (2.5.5). A *kinematic symmetry* is a time-dependent vector field X on C such that

$$X_a\frac{\partial T}{\partial q_a} + W_a\frac{\partial T}{\partial v_a} = 0 \tag{2.5.39}$$

where $W_a = v_b\,\partial X_a/\partial q_b + \partial X_a/\partial t$.

This definition is not standard.

The momentum conjugate to a kinematic symmetry is the function $p: PT \to \mathbb{R}$, where

$$p = X_a \frac{\partial T}{\partial v_a}. \tag{2.5.40}$$

This does not involve the dynamics of the system: the definition makes no mention of the forces acting on the particles.

If X is a kinematic symmetry, then the same calculation as in the proof of Noether's theorem (but using now eqn (2.3.23) rather than Lagrange's equations) shows that

$$\frac{\mathrm{d}p}{\mathrm{d}t} = X_a F_a \tag{2.5.41}$$

where p is the conjugate momentum. In other words, a kinematic symmetry gives rise to a *momentum principle* in the form of an equation: rate of change of momentum ($\dot{p}$) equals component of applied force ($X_a F_a$).

Example (2.5.3). Refer the motion of the system of particles to an inertial frame R and write $T = \frac{1}{2} \sum m_\alpha \boldsymbol{v}_\alpha \cdot \boldsymbol{v}_\alpha$. Consider the vector fields X for which

(1) $\boldsymbol{X}_\alpha = \boldsymbol{x}$, where $\boldsymbol{x}$ is a constant vector.
(2) $\boldsymbol{X}_\alpha = \boldsymbol{\Omega} \wedge \boldsymbol{r}_\alpha$, where $\boldsymbol{\Omega}$ is a constant vector.
(3) $\boldsymbol{X}_\alpha = \boldsymbol{\Omega} \wedge (\boldsymbol{r}_\alpha - \boldsymbol{c}(t))$, where $\boldsymbol{\Omega}$ is a constant vector and $\boldsymbol{c}(t)$ is the position vector of a moving point C.

In the first case, the displacements involve the translation of the whole system through $\varepsilon\boldsymbol{x}$. The $\boldsymbol{W}_\alpha$ vanish and so T is unchanged; X is therefore a kinematic symmetry. We have

$$p = \boldsymbol{x} \cdot \sum_\alpha m_\alpha \boldsymbol{v}_\alpha, \qquad X_a F_a = \boldsymbol{x} \cdot \sum \boldsymbol{F}_\alpha. \tag{2.5.42}$$

Equation (2.5.41) gives $\dot{p} = X_a F_a$. Since this is true for every choice of $\boldsymbol{x}$, we have

$$\dot{\boldsymbol{p}} = \sum \boldsymbol{F}_\alpha, \tag{2.5.43}$$

where $\boldsymbol{p} = \sum m_\alpha \boldsymbol{v}_\alpha$ is the *total linear momentum*. This is the *principle of linear momentum.*

In the second case, the displacements involve the rotation of the whole system about an axis through the origin parallel to $\boldsymbol{\Omega}$. The $\boldsymbol{W}_\alpha$ are

$$\boldsymbol{W}_\alpha = \frac{\mathrm{d}\boldsymbol{X}_\alpha}{\mathrm{d}t} = \boldsymbol{\Omega} \wedge \boldsymbol{v}_\alpha. \tag{2.5.44}$$

So the change in T under the corresponding displacement in the state of motion is

$$\delta T = \varepsilon \sum_\alpha m_\alpha \boldsymbol{v}_\alpha \,.\, \boldsymbol{W}_\alpha = \varepsilon \sum_\alpha m_\alpha \boldsymbol{v}_\alpha \,.\, (\boldsymbol{\Omega} \wedge \boldsymbol{v}_\alpha) = 0. \qquad (2.5.45)$$

Therefore X is again a kinematic symmetry. In this case we have

$$p = \boldsymbol{\Omega} \,.\, \sum m_\alpha \boldsymbol{r}_\alpha \wedge \boldsymbol{v}_\alpha,\ X_a F_a = \boldsymbol{\Omega} \,.\, \sum \boldsymbol{r}_\alpha \wedge \boldsymbol{F}_\alpha \qquad (2.5.46)$$

by the same calculation as in example (2.5.2). Equation (2.5.46) (for different values of $\boldsymbol{\Omega}$) leads to the *principle of angular momentum*

$$\dot{\boldsymbol{M}}_O = \sum_\alpha \boldsymbol{r}_\alpha \wedge \boldsymbol{F}_\alpha, \qquad (2.5.47)$$

where

$$\boldsymbol{M}_O = \sum_\alpha m_\alpha \boldsymbol{r}_\alpha \wedge \boldsymbol{v}_\alpha \qquad (2.5.48)$$

is the *total angular momentum* about the origin. The right-hand side of eqn (2.5.47) is the total moment of the forces on the particles.

In the third case,

$$\boldsymbol{W}_\alpha = \boldsymbol{\Omega} \wedge (\boldsymbol{v}_\alpha - \dot{\boldsymbol{c}}) \qquad (2.5.49)$$

and

$$\begin{aligned} \delta T &= \varepsilon \sum_\alpha m_\alpha [\boldsymbol{\Omega} \wedge (\boldsymbol{v}_\alpha - \dot{\boldsymbol{c}})] \,.\, \boldsymbol{v}_\alpha \\ &= -\varepsilon (\boldsymbol{\Omega} \wedge \dot{\boldsymbol{c}}) \,.\, \boldsymbol{p}. \end{aligned} \qquad (2.5.50)$$

In general the right-hand side does not vanish. Two important exceptions are: first, C is fixed in the inertial frame. Then $\dot{\boldsymbol{c}} = 0$ and we learn nothing new: by taking C as the origin, we come back to the second case.

The second exception is when C is the *centre of mass,* which is defined by

$$\sum_\alpha m_\alpha \boldsymbol{r}_\alpha = m\boldsymbol{c} \quad \text{where} \quad m = \sum_\alpha m_\alpha. \qquad (2.5.51)$$

Then $\dot{\boldsymbol{c}}$ is parallel to $\boldsymbol{p}$ and the right-hand side of eqn (2.5.50) vanishes. In this case, we have the *principle of angular momentum about the centre of mass*

$$\dot{\boldsymbol{M}}_{\mathrm{CM}} = \sum_\alpha (\boldsymbol{r}_\alpha - \boldsymbol{c}) \wedge \boldsymbol{F}_\alpha, \qquad (2.5.52)$$

where

$$\boldsymbol{M}_{\mathrm{CM}} = \sum_\alpha m_\alpha (\boldsymbol{r}_\alpha - \boldsymbol{c}) \wedge \boldsymbol{v}_\alpha = \boldsymbol{M}_O - \boldsymbol{c} \wedge \boldsymbol{p} \qquad (2.5.53)$$

is the total *angular momentum about the centre of mass.* The right hand side of eqn (2.5.52) is the total moment of the forces about the centre of mass.

The forces on a system of particles can usually be split into a sum

$$\boldsymbol{F}_\alpha = \boldsymbol{E}_\alpha + \boldsymbol{I}_\alpha \tag{2.5.54}$$

where the $\boldsymbol{E}_\alpha$ are *external forces* (gravity and so on) and the $\boldsymbol{I}_\alpha$ are *internal forces* that arise from the mutual interactions between the particles. The only restriction on the internal forces is that they should do no work under small instantaneous displacements that do not alter the *relative* positions of the particles; that is, under rotations and translations of the whole system. For example, the forces between the particles in a rigid body are internal in this sense: in the absence of gravity and other external forces, one cannot extract energy from a rock simply by moving it or rotating it. (If one could, the energy crisis—or, more accurately, the entropy crisis—would be over for good.)

In all three cases, only the external forces contribute to $X_a F_a$. Therefore we obtain finally

(1) Linear momentum: $\dot{\boldsymbol{p}} = \sum \boldsymbol{E}_\alpha$.
(2) Angular momentum: $\dot{\boldsymbol{M}}_O = \sum \boldsymbol{r}_\alpha \wedge \boldsymbol{E}_\alpha$.
(3) Angular momentum about the centre of mass:

$$\dot{\boldsymbol{M}}_{\mathrm{CM}} = \sum (\boldsymbol{r}_\alpha - \boldsymbol{c}) \wedge \boldsymbol{E}_\alpha,$$

where $\boldsymbol{c}$ is the position vector of the centre of mass.

It is very important to remember that the equality between rate of change of angular momentum and total moment of external forces only holds in general for *either* a fixed point in an inertial frame *or* for the centre of mass.

Exercises

(2.5.1) Spherical polar coordinates are defined by $q_1' = r$, $q_2' = \theta$, $q_1' = \varphi$, where

$$q_1 = x = r \sin\theta \cos\varphi$$
$$q_2 = y = r \sin\theta \sin\varphi$$
$$q_3 = z = r \cos\theta.$$

Show that the vector field X with q-components

$$X_1 = 0, \qquad X_2 = -q_3, \qquad X_3 = q_2$$

has q'-components

$$X_1' = 0, \qquad X_2' = -\sin\varphi, \qquad X_3' = -\cos\varphi \cot\theta.$$

Hence write down an expression in terms of the q'_a for the x-component of the angular momentum of a free particle.

(2.5.2) An astronaut is floating in empty space at rest relative to an inertial frame with her arms by her side. Explain how it is that by waving her arms and then returning them to their original position, she can rotate her body, but cannot move her centre of mass.

(2.5.3) In example (2.5.3): show that if $\boldsymbol{F}_\alpha = m_\alpha \boldsymbol{g}$ where $\boldsymbol{g}$ is constant, then $\sum \boldsymbol{F}_\alpha = m\boldsymbol{g}$ and $\sum \boldsymbol{r}_\alpha \wedge \boldsymbol{F}_\alpha = m\boldsymbol{c} \wedge \boldsymbol{g}$, where $\boldsymbol{c}$ is the position vector of the centre of mass. Deduce that the effect of a uniform gravitational field on the total linear momentum and angular momentum is the same as that of a single force $m\boldsymbol{g}$ acting through the centre of mass. Show by counter-example that this is not true for a non-uniform field.

(2.5.4) Show that if X and $\hat{X}$ are time-dependent vector fields on C, then so is $[X, \hat{X}]$, where $[X, \hat{X}]$ has components

$$X_b \frac{\partial \hat{X}_a}{\partial q_b} - \hat{X}_b \frac{\partial X_a}{\partial q_b}.$$

Show that if X and $\hat{X}$ are dynamical symmetries of a system with Lagrangian L, then so is $[X, \hat{X}]$.

For a free particle, X and $\hat{X}$ generate, respectively, small rotations about the z and x axes. What displacements are generated by $[X, \hat{X}]$?

(2.5.5)* Show that under the small displacements in PT generated by a time-dependent vector field X on C, the change in $\partial L/\partial v_a$ is

$$\delta\left(\frac{\partial L}{\partial v_a}\right) = \varepsilon X_b \frac{\partial^2 L}{\partial v_a\, \partial q_b} + \varepsilon W_b \frac{\partial^2 L}{\partial v_a\, \partial v_b}.$$

Show that if X is a dynamical symmetry, then

$$\delta\left(\frac{\partial L}{\partial v_a}\right) = -\varepsilon \frac{\partial X_b}{\partial q_a} \frac{\partial L}{\partial v_b}$$

and

$$\delta\left(\frac{\partial L}{\partial q_a}\right) = -\varepsilon \frac{\partial X_b}{\partial q_a} \frac{\partial L}{\partial q_b} - \varepsilon \frac{\partial W_b}{\partial q_a} \frac{\partial L}{\partial v_b}.$$

Show that

$$\frac{\mathrm{d}}{\mathrm{d}t}\left(\frac{\partial X_b}{\partial q_a}\right) = \frac{\partial W_b}{\partial q_a}.$$

Hence show that if X is a dynamical symmetry and $q_a = x_a(t)$ is a solution of Lagrange's equations, then so is $q_a = x_a(t) + \varepsilon X_a(x(t), t)$ (to the first order in ε).

2.6* The calculus of variations

Lagrange's equations also arise in the calculus of variations. In this section, we shall look at the main ideas involved in the analysis of variational problems and at their connection with mechanics through Hamilton's principle; we shall see how they can be used to give an alternative proof of proposition (2.3.1).

Let $F = F(q, v, t)$ be a function on the time–phase space PT of a system with n degrees of freedom (PT could also be the constrained phase space of a system subject to holonomic constraints); and let γ be a curve in CT joining the point A, which has coordinates (A_a, t_1), to the point B, which has coordinates (B_a, t_2).

If t is used as a parameter along γ, then $q_a = g_a(t)$, where $g_1, \ldots, g_n$ are functions of t. Put

$$J_F(\gamma) = \int_{t_1}^{t_2} F(g(t), \dot{g}(t), t)\, \mathrm{d}t. \tag{2.6.1}$$

The integrand is F evaluated at points of the extension of γ to PT. Since the extension of a curve is the same irrespective of the choice of coordinates, the value of the integral depends only on F and γ, and not on the choice of coordinate system.

We shall think of J_F as a real-valued function (or 'functional') on the set of curves joining A to B.

In section 2.3, the equation of a curve in CT was written as $q_a = q_a(t)$. We have now reached the point at which this abuse of notation would lead irrevocably to the *first fundamental confusion of calculus,* which arises when the same symbol is used for a function as for its value, as in $y = y(x)$ (interesting examples can be created by taking 'y' to be 'cos'). We shall avoid the pitfalls of this confusion by using systematically—as we have already done occasionally—a notation that distinguishes q_a (a coordinate) from g_a (a function of time).

By using t as a parameter, we are restricting the class of curves that we can consider: for example, when $n = 1$, the curve $t = q^2$ is inadmissible. However, the restriction has no significant consequences.

The calculus of variations is concerned with the problem of finding the *critical curves* of J_F: the curves γ for which $J_F(\gamma)$ is unchanged under small deformations of γ. It is analogous to the problem of finding the critical points of a function $f(x, y)$—the local maxima, minima, and saddle points of f. These are the points at which the value of f is unchanged to the first order under small variations in the values of the coordinates x and y. They are characterized by the vanishing of grad f.

* Section contains harder material that can be omitted.

Definition (2.6.1). A *variation* or *deformation* of γ is a family γ_s of curves in CT labelled by $s \in \mathbb{R}$ such that:
(i) for each fixed value of s, γ_s is a curve in CT joining A to B;
(ii) $\gamma_0 = \gamma$.

In generalized coordinates, a deformation is defined by giving $q_a = G_a(s, t)$ as a function of the two variables t and s: for fixed s, $q_a = G_a(s, t)$ is the equation of γ_s. Condition (i) translates into

$$G_a(s, t_1) = A_a, \qquad G_a(s, t_2) = B_a \tag{2.6.2}$$

and condition (ii) translates into the equation $G_a(0, t) = g_a(t)$ for all $t \in [t_1, t_2]$ (Fig. 2.6.1).

Let γ_s be a deformation of γ and, for each s, put

$$J(s) = J_F(\gamma_s). \tag{2.6.3}$$

Then $J(0) = J_F(\gamma)$; and the derivative $J'(0)$ measures the rate of change in J_F as we move from γ to γ_s for small values of s.

Proposition (2.6.1). Let $X_a(t) = \partial G_a/\partial s\,|_{s=0}$. Then

$$J'(0) = \int_{t_1}^{t_2} Q_a(t) X_a(t)\,\mathrm{d}t \tag{2.6.4}$$

where

$$Q_a(t) = -\frac{\mathrm{d}}{\mathrm{d}t}\left(\frac{\partial F}{\partial v_a}\right) + \frac{\partial F}{\partial q_a}, \tag{2.6.5}$$

the right-hand side being evaluated along the extension of γ; that is, with $q_a = g_a(t)$ and $v_a = \dot{g}_a(t)$.

Proof. For small values of s,

$$G_a(s, t) = g_a(t) + sX_a(t) + \mathrm{O}(s^2), \tag{2.6.6}$$

where, by condition (i), $X_a(t_1) = 0 = X_a(t_2)$. Therefore

$$\begin{aligned} J(s) &= \int_{t_1}^{t_2} F(g(t) + sX(t), \dot{g}(t) + s\dot{X}(t), t)\,\mathrm{d}t + \mathrm{O}(s^2) \\ &= J(0) + \int_{t_1}^{t_2} s\left(\frac{\partial F}{\partial q_a} X_a + \frac{\partial F}{\partial v_a}\dot{X}_a\right)\mathrm{d}t + \mathrm{O}(s^2). \end{aligned} \tag{2.6.7}$$

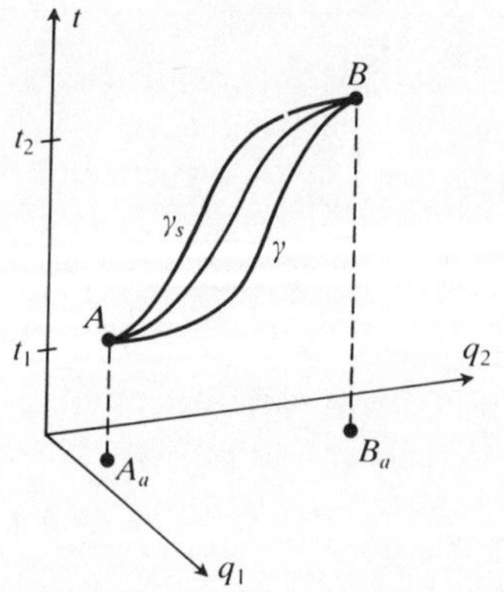

Fig. 2.6.1 A variation of a curve γ in CT.

Hence, by integrating by parts,

$$\begin{aligned} J'(0) &= \int_{t_1}^{t_2} \left(\frac{\partial F}{\partial q_a} X_a + \frac{\partial F}{\partial v_a} \dot{X}_a \right) \mathrm{d}t \\ &= \int_{t_1}^{t_2} \left[\frac{\partial F}{\partial q_a} - \frac{\mathrm{d}}{\mathrm{d}t} \left(\frac{\partial F}{\partial v_a} \right) \right] X_a \, \mathrm{d}t \end{aligned} \tag{2.6.8}$$

since $X_a(t_1) = 0 = X_a(t_2)$. The integrand is interpreted according to the usual rules: first calculate the partial derivatives of F with respect to q_a and v_a; then substitute $q_a = g_a(t)$ and $v_a = \dot{g}_a(t)$, and take the time derivative $\mathrm{d}/\mathrm{d}t$.

□

In order to make sense of the calculations in the proof, we should impose differentiability conditions on the functions involved. In the most straightforward approach, one requires that the g_a should be twice continuously differentiable with respect to t on $[t_1, t_2]$; that the G_a should be twice continuously differentiable with respect to s and t on $[-\varepsilon, \varepsilon] \times [t_1, t_2]$ for some $\varepsilon > 0$; and that F should be twice continuously differentiable with respect to all its arguments. However, proposition (2.6.2) can in fact be proved under weaker conditions.

Definition (2.6.2). A curve γ from A to B in CT is a *critical curve* of J_F if $J'(0) = 0$ for every deformation of γ.

For a critical curve, the value of J_F is unchanged to the first order in s under small deformations.

Proposition (2.6.2) Euler's theorem. If γ is a critical curve then

$$\frac{\mathrm{d}}{\mathrm{d}t}\left(\frac{\partial F}{\partial v_a}\right)-\frac{\partial F}{\partial q_a}=0 \tag{2.6.9}$$

along the extension of γ.

Proof. Suppose that γ is critical. Given any set of (twice continuously differentiable) functions $X_a(t)$ such that $X_a(t_1)=X_a(t_2)=0$, we can find a deformation of γ such that eqn (2.6.6) holds; for example we can take

$$G_a(s, t)=g_a(t)+sX_a(t). \tag{2.6.10}$$

Hence, for any such $X_a(t)$,

$$\int_{t_1}^{t_2} Q_a(t)X_a(t)\,\mathrm{d}t=0. \tag{2.6.11}$$

Suppose that $Q_1(t_3)>0$ for some $t_3\in(t_1, t_2)$. Then, by continuity, there exists $\varepsilon>0$ such that $t_3-\varepsilon>t_1$, $t_3+\varepsilon<t_2$, and $Q_1(t)>0$ whenever $t_3-\varepsilon<t<t_3+\varepsilon$.

Put

$$X_1(t)=\begin{cases}(t_3+\varepsilon-t)^3(t-t_3+\varepsilon)^3 & t_3-\varepsilon<t<t_3+\varepsilon\\ 0 & \text{otherwise}\end{cases} \tag{2.6.12}$$

$$X_2(t)=X_3(t)=\cdots=X_n(t)=0.$$

Then

$$\int_{t_1}^{t_2} Q_aX_a\,\mathrm{d}t=\int_{t_3-\varepsilon}^{t_3+\varepsilon} Q_1X_1\,\mathrm{d}t>0. \tag{2.6.13}$$

But the X_a are twice continuously differentiable and they vanish at t_1 and t_2. Hence eqn (2.6.13) contradicts eqn (2.6.11). A similar argument disposes of $Q_1(t_3)<0$. Therefore, by continuity, $Q_1(t)=0$ for all $t\in[t_1, t_2]$ and similarly $Q_a(t)=0$ for the other values of a. □

Example (2.6.1). With $n=1$: suppose that $F=\frac{1}{2}v^2/q^2$, $A=(1, 0)$, and $B=(\mathrm{e}, 1)$. Then eqn (2.6.9) is $q\ddot{q}=\dot{q}^2$, which has the general solution $q=\alpha\,\mathrm{e}^{\beta t}$ for constant α and β. The critical curve joining A to B has $q=1$ when $t=0$ and $q=\mathrm{e}$ when $t=1$: it is $q=\mathrm{e}^t$.

As an immediate corollary, we have the following.

Proposition (2.6.3) Hamilton's principle. The orbits in CT of a conservative system with Lagrangian L are the critical curves of J_L.

Hamilton's principle gives an elegant way of picking out the evolution of a system from a specified initial configuration at time t_1 to a specified final configuration at time t_2. It would be pleasing if one could say that the orbits actually minimized J_L; but, unfortunately this is only true if the time interval $t_2 - t_1$ is small and the initial configuration is close to the final one.

Finally, we can use proposition (2.6.1) to prove proposition (2.3.1). Suppose that we have two systems of generalized coordinates q_a and $\tilde{q}_a$ (as in section 2.3); and that we have a curve γ from A to B, as above. Consider a deformation of γ and suppose that it is given by

$q_a = G_a(s, t)$ in the coordinates q_a; and by
$\tilde{q}_a = \tilde{G}_a(s, t)$ in the coordinates $\tilde{q}_a$.

Then

$$J'(0) = \int_{t_1}^{t_2} Q_a X_a \, \mathrm{d}t = \int_{t_1}^{t_2} \tilde{Q}_a \tilde{X}_a \, \mathrm{d}t \tag{2.6.14}$$

where Q_a and X_a are as before,

$$\tilde{X}_a = \left.\frac{\partial \tilde{G}_a}{\partial s}\right|_{s=0} \quad \text{and} \quad \tilde{Q}_a = -\frac{\mathrm{d}}{\mathrm{d}t}\left(\frac{\partial F}{\partial \tilde{v}_a}\right) + \frac{\partial F}{\partial \tilde{q}_a}. \tag{2.6.15}$$

But $\tilde{X}_a = X_b \partial \tilde{q}_a / \partial q_b$ since

$$\frac{\partial \tilde{G}_a}{\partial s} = \frac{\partial \tilde{q}_a}{\partial q_b}\frac{\partial G_b}{\partial s} \tag{2.6.16}$$

by the chain rule. Therefore

$$\int_{t_1}^{t_2} \left[Q_a - \tilde{Q}_b \frac{\partial \tilde{q}_b}{\partial q_a}\right] X_a \, \mathrm{d}t = 0 \tag{2.6.17}$$

and so

$$\frac{\mathrm{d}}{\mathrm{d}t}\left(\frac{\partial F}{\partial v_a}\right) - \frac{\partial F}{\partial q_a} = \frac{\partial \tilde{q}_b}{\partial q_a}\left[\frac{\mathrm{d}}{\mathrm{d}t}\left(\frac{\partial F}{\partial \tilde{v}_b}\right) - \frac{\partial F}{\partial \tilde{q}_b}\right] \tag{2.6.18}$$

by the same argument as in the proof of Euler's theorem. □

I am grateful to Handel Davies for suggesting the inclusion of this argument.

Exercises

(2.6.1) With $n = 1$: suppose that $F = v^2$, $A = (0, 0)$, and $B = (1, 1)$. Show that the critical curve γ is given by $q = t$. Show that $G(s, t) = t + st(t-1)$ defines a deformation of γ. Show that $J(s) = \frac{1}{3}(s^2 + 3)$ and so check that $J(s)$ is minimal when $s = 0$.

(2.6.2)* Establish the result of exercise (2.5.5) by a variational argument.

(2.6.3) Show that if every curve from A to B is a critical curve of J_F, then

$$F(q, v, t) = \frac{\partial f}{\partial q_a} v_a + \frac{\partial f}{\partial t}$$

for some function $f = f(q, t)$ on CT.

(2.6.4) A particle of unit mass is constrained to move on the surface of a unit sphere, but is otherwise free. Show that the orbits are great circles traversed at uniform speeds. Show that if γ is a complete circuit of a great circle in time t, then

$$J_L(\gamma) = 2\pi^2/t.$$

Does γ minimize J_L over all curves on the sphere that start and end at a point P on the equator and take time t for the round trip from P back to P?

3 Rigid bodies

3.1 Kinetic energy and the inertia matrix

The motion of a rigid body at any instant can be characterized by the six components of two vectors: the angular velocity $\boldsymbol{\omega}$ and the velocity $\boldsymbol{v}$ of a chosen point O of the body. There are therefore six degrees of freedom and we should be able to describe the evolution by introducing six generalized coordinates: three for position and three for orientation.

The position coordinates are straightforward: we can use the three Cartesian coordinates of O in some inertial frame. The components of $\boldsymbol{v}$ are then the corresponding generalized velocities. Unfortunately, as we shall see in section 3.3, it is a more complicated problem to find convenient coordinates to describe the rotational degrees of freedom; and it is, in fact, impossible to find three generalized coordinates for which the components of $\boldsymbol{\omega}$ are the corresponding generalized velocities. Whatever angular coordinates are used, the task of expressing $\boldsymbol{\omega}$ in terms of their time derivatives is always a source of complication.

Having chosen coordinates $q_1, \ldots, q_6$, we must express the kinetic energy as a function of q_a, v_a, and t. We can then write down the equations of motion in terms of the q-components of the external forces. In doing this, we shall think of the body as a collection of a large number of particles, with position vectors $\boldsymbol{r}_\alpha$, subject to the constraints

$$|\boldsymbol{r}_\alpha - \boldsymbol{r}_\beta| = \text{constant}. \qquad (3.1.1)$$

The constraint forces are the forces between the particles and we assume that they do work during instantaneous displacements of the body which leave unchanged the relative positions of the individual particles; they therefore make no contribution to the equations of motion for the coordinates q_a.

This model of a rigid body is not realistic; and, in any case, in applications one usually thinks of a rigid body as a continuous distribution of matter, rather than as a large, but finite, collection of particles. The derivation of the equations of motion is therefore heuristic: one should regard the rigid body equations (for example, in the vector form given in section 3.2) as fresh axioms of mechanics, rather than as direct consequences of Newton's laws.

Before turning to the details of the coordinate problem, we shall consider the kinetic energy of a rigid body and its expression in terms of velocity and angular velocity.

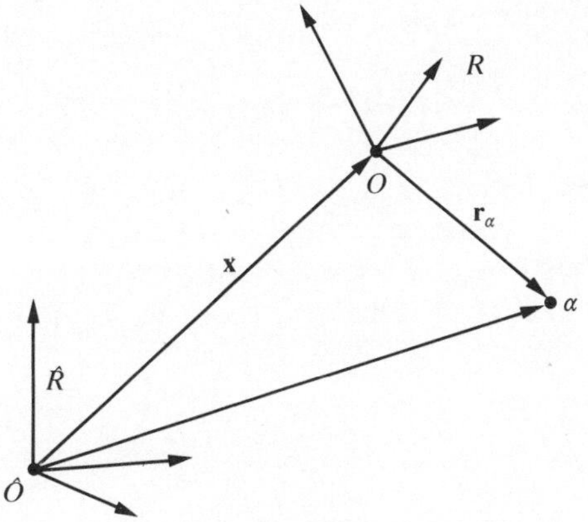

Fig. 3.1.1

Let $\hat{R} = (\hat{O}, \hat{B})$ be an inertial frame; and let $R = (O, B)$ be a rest frame of the body (Fig. 3.1.1). Suppose that particle α of the body has O-position vector $\boldsymbol{r}_\alpha$. Then its $\hat{O}$-position vector is $\boldsymbol{r}_\alpha + \boldsymbol{x}$, where $\boldsymbol{x}$ is the vector from $\hat{O}$ to O. Hence its velocity relative to $\hat{R}$ is

$$\hat{\boldsymbol{v}}_\alpha = \dot{\boldsymbol{x}} + \boldsymbol{\omega} \wedge \boldsymbol{r}_\alpha, \tag{3.1.2}$$

where $\boldsymbol{\omega}$ is the angular velocity of the body and the dot denotes the time derivative with respect to $\hat{B}$. It follows that the total kinetic energy (relative to the inertial frame) is

$$T = \sum \tfrac{1}{2} m_\alpha \hat{\boldsymbol{v}}_\alpha \cdot \hat{\boldsymbol{v}}_\alpha$$

$$= \tfrac{1}{2} m \dot{\boldsymbol{x}} \cdot \dot{\boldsymbol{x}} + \dot{\boldsymbol{x}} \cdot \left(\boldsymbol{\omega} \wedge \sum m_\alpha \boldsymbol{r}_\alpha\right) + \tfrac{1}{2} \sum m_\alpha (\boldsymbol{\omega} \wedge \boldsymbol{r}_\alpha) \cdot (\boldsymbol{\omega} \wedge \boldsymbol{r}_\alpha), \tag{3.1.3}$$

where the sum is over all the particles and $m = \sum m_\alpha$ is the total mass.

The second term can be simplified to

$$m \dot{\boldsymbol{x}} \cdot (\boldsymbol{\omega} \wedge \boldsymbol{c}), \tag{3.1.4}$$

where $\boldsymbol{c}$ is the O-position vector of the centre of mass, defined by

$$m\boldsymbol{c} = \sum m_\alpha \boldsymbol{r}_\alpha. \tag{3.1.5}$$

To make better sense of the third term, we shall replace the particles by a continuous distribution of matter. Then

$$\tfrac{1}{2} \sum m_\alpha (\boldsymbol{\omega} \wedge \boldsymbol{r}_\alpha) \cdot (\boldsymbol{\omega} \wedge \boldsymbol{r}_\alpha) \tag{3.1.6}$$

becomes an integral over the body

$$\frac{1}{2}\int \rho(\boldsymbol{\omega}\wedge \boldsymbol{r}).(\boldsymbol{\omega}\wedge \boldsymbol{r})\,\mathrm{d}\tau = \frac{1}{2}\int \rho[(\boldsymbol{\omega}.\boldsymbol{\omega})(\boldsymbol{r}.\boldsymbol{r}) - (\boldsymbol{\omega}.\boldsymbol{r})^2]\,\mathrm{d}\tau, \quad (3.1.7)$$

in which ρ is the density, $\boldsymbol{r}$ is the O-position of a general point of the body, and $\mathrm{d}\tau$ is the volume element.

In general, the density ρ depends on $\boldsymbol{r}$; but $\boldsymbol{\omega}$, which is simply the angular velocity of B relative to $\hat{B}$, is the same for all particles at any one time. Therefore the integral (3.1.7) is equal to

$$\tfrac{1}{2}\omega_i\omega_j \int \rho(r_k r_k \delta_{ij} - r_i r_j)\,\mathrm{d}\tau, \quad (3.1.8)$$

where the ω_i are the B-components of $\boldsymbol{\omega}$ and the r_i are the B-components of $\boldsymbol{r}$.

Definition (3.1.1). The *inertia matrix* of the body in the rest frame R is a 3×3 symmetric matrix $J(R)$ with entries

$$J_{ij} = \int \rho(r_k r_k \delta_{ij} - r_i r_j)\,\mathrm{d}\tau. \quad (3.1.9)$$

The notation is, perhaps, less than transparent. To bring it into a more familiar form, write x, y, and z for r_1, r_2, and r_3, and put

$$\begin{aligned} A &= \int \rho(y^2+z^2)\,\mathrm{d}\tau & F &= \int \rho yz\,\mathrm{d}\tau \\ B &= \int \rho(x^2+z^2)\,\mathrm{d}\tau & G &= \int \rho xz\,\mathrm{d}\tau \\ C &= \int \rho(x^2+y^2)\,\mathrm{d}\tau & H &= \int \rho xy\,\mathrm{d}\tau. \end{aligned} \quad (3.1.10)$$

Then

$$J(R) = \begin{pmatrix} A & -H & -G \\ -H & B & -F \\ -G & -F & C \end{pmatrix}. \quad (3.1.11)$$

Definition (3.1.2). The diagonal entries A, B, and C are the *moments of inertia* about the x, y, and z coordinate axes; F, G, and H are the *products of inertia* with respect to the yz, zx, and xy coordinate planes.

We can now write

$$T = \tfrac{1}{2}m\dot{\boldsymbol{x}}\,.\,\dot{\boldsymbol{x}} + m\dot{\boldsymbol{x}}\,.\,(\boldsymbol{\omega} \wedge \boldsymbol{c}) + \tfrac{1}{2}J_{ij}\omega_i\omega_j. \tag{3.1.12}$$

The rest frame R is not unique and a good choice can simplify matters. The following propositions determine the relationship between the inertia matrices of the body in different rest frames.

Proposition (3.1.1) Parallel axes theorem. Let O' be the centre of mass and let $R = (O, B)$ and $R' = (O', B)$ be two rest frames with parallel coordinate axes, R' having origin O'. Let J_{ij} and J'_{ij} be the entries in the inertia matrices $J(R)$ and $J(R')$. Then

$$J_{ij} = J'_{ij} + m(c_kc_k\delta_{ij} - c_ic_j) \tag{3.1.13}$$

where $\boldsymbol{c}$ is the vector from O to O' and the c_i are the B-components of $\boldsymbol{c}$.

Proof. If $\boldsymbol{r}$ is the O-position vector of a point of the body, then its O'-position vector is $\boldsymbol{s}$, where $\boldsymbol{r} = \boldsymbol{s} + \boldsymbol{c}$. Since O' is the centre of mass

$$\int \rho c_i s_j \,\mathrm{d}\tau = c_i \int \rho s_j \,\mathrm{d}\tau = 0. \tag{3.1.14}$$

Therefore, by substituting $r_i = s_i + c_i$,

$$\begin{aligned} J_{ij} &= \int \rho[(s_k + c_k)(s_k + c_k)\delta_{ij} - (s_i + c_i)(s_j + c_j)]\,\mathrm{d}\tau \\ &= \int \rho(s_ks_k\delta_{ij} - s_is_j)\,\mathrm{d}\tau + (c_kc_k\delta_{ij} - c_ic_j)\int \rho\,\mathrm{d}\tau \\ &= J'_{ij} + m(c_kc_k\delta_{ij} - c_ic_j). \end{aligned} \tag{3.1.15}$$

□

In the matrix notation, eqn (3.1.15) reads

$$J(R) = J(R') + m[(C^{\mathrm{t}}C)I - CC^{\mathrm{t}}] \tag{3.1.16}$$

where C is the column vector with entries c_1, c_2, c_3 and I is the identity matrix. Note that $C^{\mathrm{t}}C = c_kc_k = \boldsymbol{c}\,.\,\boldsymbol{c}$ and that CC^{t} is the 3×3 matrix with entries c_ic_j.

The parallel axes theorem states that the inertia matrix in a rest frame R is the sum of two terms: the inertia matrix of a single particle of mass m at the centre of mass (the expression with square brackets in eqn (3.1.16)) and the inertia matrix of the body in a frame with axes parallel to those of R and origin at the centre of mass. For the diagonal entries, the theorem reduces to the familiar parallel axis theorem for moments of inertia.

Proposition (3.1.2) The tensor property. Let $R = (O, B)$ and $R' = (O, B')$ be two rest frames with the same origin and let H be the transition matrix from B' to B. Let J_{ij} and J'_{ij} be the entries in the inertia matrices $J(R)$ and $J(R')$. Then

$$J_{ij} = H_{ip} J'_{pq} H_{jq}. \tag{3.1.17}$$

Proof. Let $\boldsymbol{r}$ be the O-position vector of a point of the body. Then the B and B' components of $\boldsymbol{r}$ are related by $r_i = H_{ip} r'_p$. Hence

$$\begin{aligned} J_{ij} &= \int \rho (r_k r_k \delta_{ij} - r_i r_j)\,\mathrm{d}\tau \\ &= \int \rho H_{ip} H_{jq} (r'_k r'_k \delta_{pq} - r'_p r'_q)\,\mathrm{d}\tau \\ &= H_{ip} J'_{pq} H_{jq} \end{aligned} \tag{3.1.18}$$

since $r_k r_k = r'_k r'_k$ and $H_{ip} H_{jq} \delta_{pq} = H_{ip} H_{jp} = \delta_{ij}$. □

In matrix notation, eqn (3.1.17) reads

$$J(R) = H J(R') H^{t}, \tag{3.1.19}$$

which is equivalent to $J(R') = H^{t} J(R) H$. But for any symmetric matrix J, there exists an orthogonal matrix H such that $H^{t} J H$ is diagonal. Given O, therefore, it is possible to choose the orthonormal triad B' so that, with $R' = (O, B')$,

$$J(R') = \begin{pmatrix} J'_{11} & 0 & 0 \\ 0 & J'_{22} & 0 \\ 0 & 0 & J'_{33} \end{pmatrix}. \tag{3.1.20}$$

Definition (3.1.3). Let $J(R)$ be the inertia matrix in a rest frame $R = (O, B)$. A *principal axis* at O is a line through O in the direction of an eigenvector of $J(R)$; that is, in the direction of vector $\boldsymbol{x}$ such that

$$J_{ij} x_j = \lambda x_i, \quad \text{or} \quad J(R) \begin{pmatrix} x_1 \\ x_2 \\ x_3 \end{pmatrix} = \lambda \begin{pmatrix} x_1 \\ x_2 \\ x_3 \end{pmatrix}, \tag{3.1.21}$$

where $\lambda \in \mathbb{R}$. The eigenvalue λ is the corresponding *principal moment of inertia*.

The principal axes and principal moments of inertia depend on O, but not on the choice of B. If B' is a second orthonormal triad and H is the transition matrix from B' to B, then the B'-components of $\boldsymbol{x}$ are $x'_j = x_i H_{ij}$. Hence if $\boldsymbol{x}$ satisfies eqn (3.1.21), then

$$J'_{ij}x'_j = H_{pi}J_{pq}H_{qj}x'_j = H_{pi}J_{pq}x_q = \lambda x'_i. \tag{3.1.22}$$

It follows that the line through O in the direction of $\boldsymbol{x}$ still satisfies the definition when B is replaced by B'.

Some familiar results from linear algebra translate into statements about principal axes and principal moments. For example, two eigenvectors of a symmetric matrix corresponding to distinct eigenvalues are necessarily orthogonal: this becomes the statement that the principal axes at O of distinct principal moments of inertia are orthogonal. If two vectors are eigenvectors of a matrix with the same eigenvalue, then any linear combination of the two is also an eigenvector with this eigenvalue: thus if two lines through O are principal axes corresponding to the same principal moment of inertia λ, then every line through O in the plane that they span is also a principal axis with principal moment of inertia λ.

If the triad $B' = (\boldsymbol{e}'_1, \boldsymbol{e}'_2, \boldsymbol{e}'_3)$ in $R' = (O, B')$ is chosen so that $J(R')$ is diagonal then the coordinate axes will be principal axes and the diagonal entries in $J(R')$ will be the corresponding principal moments of inertia. Conversely, if one takes three orthogonal principal axes at O as coordinate axes, then the inertia matrix will be diagonal, with the principal moments of inertia as diagonal entries.

Example (3.1.1).† The moment of inertia of a uniform spherical shell of mass m and radius a about any axis through its centre is $\frac{2}{3}ma^2$. Hence the inertia matrix at the centre is $\frac{2}{3}ma^2I$ (irrespective of the choice of triad).

The upper and lower hemispheres contribute equally to the integrals for the moments and products of inertia. Therefore the inertia matrix of a uniform hemispherical shell of mass m and radius a at its centre O is also $\frac{2}{3}ma^2I$ (half the matrix of a spherical shell of mass $2m$).

We shall use the parallel axes theorem to find the inertia matrix at a point P on the rim of the hemisphere. Since neither O nor P is the centre of mass, we must do this in two steps.

Let $B = (\boldsymbol{i}, \boldsymbol{j}, \boldsymbol{k})$, where $\boldsymbol{i}$ and $\boldsymbol{j}$ are parallel to the base, with $\boldsymbol{i}$ pointing from O to P, and $\boldsymbol{k}$ is normal to the base (Fig. 3.1.2).

The centre of mass C has O-position vector $-\frac{1}{2}a\boldsymbol{k}$. Therefore, if $R = (C, B)$, then

$$J(R) = \tfrac{2}{3}ma^2I - \tfrac{1}{4}ma^2\left[I - \begin{pmatrix} 0 & 0 & 0 \\ 0 & 0 & 0 \\ 0 & 0 & 1 \end{pmatrix}\right] \tag{3.1.23}$$

by the parallel axes theorem.

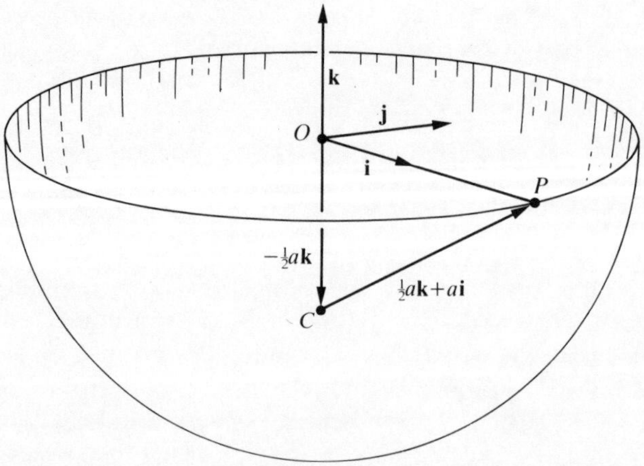

Fig. 3.1.2

The point P has C-position vector $a\boldsymbol{i} + \frac{1}{2}a\boldsymbol{k}$. Therefore, if $R' = (P, B)$, then a second application of the theorem gives

$$J(R') = J(R) + \tfrac{1}{4}ma^2\left[5I - \begin{pmatrix} 4 & 0 & 2 \\ 0 & 0 & 0 \\ 2 & 0 & 1 \end{pmatrix}\right]$$

$$= \tfrac{1}{6}ma^2\begin{pmatrix} 4 & 0 & -3 \\ 0 & 10 & 0 \\ -3 & 0 & 10 \end{pmatrix}. \qquad (3.1.24)$$

The principal moments of inertia at P are $\frac{1}{6}ma^2s$, where s is a root of

$$\begin{vmatrix} 4-s & 0 & -3 \\ 0 & 10-s & 0 \\ -3 & 0 & 10-s \end{vmatrix} = 0. \qquad (3.1.25)$$

They are

$$\tfrac{5}{3}ma^2, \quad \text{and} \quad \tfrac{1}{6}ma^2(7 \pm 3\sqrt{2}); \qquad (3.1.26)$$

and the principal axes are in the directions of the corresponding eigenvectors

$$\begin{pmatrix} 0 \\ 1 \\ 0 \end{pmatrix}, \begin{pmatrix} -1 \\ 0 \\ 1+\sqrt{2} \end{pmatrix}, \begin{pmatrix} -1 \\ 0 \\ 1-\sqrt{2} \end{pmatrix}. \qquad (3.1.27)$$

Note that the parallel axes theorem cannot be used to go directly from O to P.

□

Suppose that the rest frame $R=(O,(\boldsymbol{e}_1,\boldsymbol{e}_2,\boldsymbol{e}_3))$ has its axes aligned with principal axes at O. Then $F=G=H=0$ and

$$\tfrac{1}{2}J_{ij}\omega_i\omega_j=\tfrac{1}{2}(A\omega_1^2+B\omega_2^2+C\omega_3^2). \tag{3.1.28}$$

(From now on, $A, B, C, \ldots$ are as in eqn (3.1.11).)

The expression (3.1.12) for T simplifies further in either of two cases.

1) The origin O of the rest frame is the centre of mass. Then $\boldsymbol{c}=0$ and

$$T=\tfrac{1}{2}m\dot{\boldsymbol{x}}\,.\,\dot{\boldsymbol{x}}+\tfrac{1}{2}(A\omega_1^2+B\omega_2^2+C\omega_3^2) \tag{3.1.29}$$

(with principal axes at the centre of mass as the axes of R). Note that in this case T is the sum of two parts: the first involves only the translational degrees of freedom; the second involves only the rotational degrees of freedom.

2) The origin O is at rest relative to $\hat{R}$. This is the case in which the body is rotating about a fixed point in $\hat{R}$. We then have $\dot{\boldsymbol{x}}=0$ and

$$T=\tfrac{1}{2}(A\omega_1^2+B\omega_2^2+C\omega_3^2) \tag{3.1.30}$$

(with the principal axes at O as the R coordinate axes).

Exercises

(3.1.1) Let $R=(O,B)$ be a rest frame of a rigid body and let the entries in $J(R)$ be J_{ij}. Show that the moment of inertia about an axis through O in the direction of a unit vector with components x_i is $J_{ij}x_ix_j$.

(3.1.2) Show that the inertia matrix at the centre of a uniform solid cube with mass m and edges of length $2a$ is $\tfrac{2}{3}ma^2I$. Find the principal axes and principal moments of inertia at a vertex.

(3.1.3)† Show that the principal moments of inertia at the centre of mass of a uniform solid circular cylinder, radius a, height $2h$, and mass m, are $\tfrac{1}{2}ma^2$ and $\tfrac{1}{12}m(4h^2+3a^2)$ (repeated).

Find the principal axes and principal moments of inertia at a point distance D from the centre of mass in the plane through the centre of mass perpendicular to the axis of the cylinder.

(3.1.4) Show that the kinetic energy of a uniform rod of mass m is

$$T=\tfrac{1}{6}m(\boldsymbol{u}\,.\,\boldsymbol{u}+\boldsymbol{u}\,.\,\boldsymbol{v}+\boldsymbol{v}\,.\,\boldsymbol{v})$$

where $\boldsymbol{u}$ and $\boldsymbol{v}$ are the velocities of the two ends.

(3.1.5)† Show that if a line passing through the centre of mass of a rigid body is a principal axis at one point of the line, then it is a principal axis at every point of the line.

(3.1.6)† A rigid body has *inertial symmetry* at a point P if the principal moments of inertia at P are all equal. Show that if a body has inertial symmetry at one point, then the principal moments of inertia at the centre of mass cannot all be distinct.

A uniform solid right circular cone has height h and base of radius a. For what values of h/a does the cone have inertial symmetry at its vertex?

(3.1.7) Show that the inertia matrix at the centre of any uniform Platonic solid is a multiple of the identity matrix.

3.2 Linear and angular momentum

Before we turn to Lagrange's equations for rigid body motion, we shall consider some consequences of the principles of linear and angular momentum.

Let $\hat{R} = (\hat{O}, (\hat{\boldsymbol{e}}_1, \hat{\boldsymbol{e}}_2, \hat{\boldsymbol{e}}_3))$ be an inertial frame as in the last section, and let $R = (O, (\boldsymbol{e}_1, \boldsymbol{e}_2, \boldsymbol{e}_3))$ be a rest frame of the rigid body, with its origin O at the centre of mass of the body.

The total linear momentum relative to the inertial frame is

$$\boldsymbol{p} = m\dot{\boldsymbol{x}}, \tag{3.2.1}$$

where $\boldsymbol{x}$ is the vector from $\hat{O}$ to O and the dot denotes the time derivative with respect to $\hat{R}$. With the body represented as a continuous distribution of matter, the total angular momentum about the centre of mass is

$$\boldsymbol{M}_{\mathrm{CM}} = \int \rho \boldsymbol{r} \wedge \hat{\boldsymbol{v}}\, \mathrm{d}\tau = \int \rho \boldsymbol{r} \wedge \dot{\boldsymbol{r}}\, \mathrm{d}\tau + \left[\int \rho \boldsymbol{r}\, \mathrm{d}\tau\right] \wedge \dot{\boldsymbol{x}} = \int \rho \boldsymbol{r} \wedge \dot{\boldsymbol{r}}\, \mathrm{d}\tau, \tag{3.2.2}$$

where $\boldsymbol{r}$ is the position vector of a typical particle from O and $\hat{\boldsymbol{v}} = \dot{\boldsymbol{r}} + \dot{\boldsymbol{x}}$ is its velocity relative to $\hat{R}$ (the term in square brackets vanishes since O is the centre of mass).

By the principles of linear and angular momentum,

$$\dot{\boldsymbol{p}} = \sum \boldsymbol{E} \quad \text{and} \quad \dot{\boldsymbol{M}}_{\mathrm{CM}} = \sum \boldsymbol{r} \wedge \boldsymbol{E}. \tag{3.2.3}$$

In the first equation, $\sum \boldsymbol{E}$ is the sum of the external forces. In the second, the sum is over the points at which the external forces act.

These two vector equations determine—in principle—the evolution of the six degrees of freedom. But we cannot make very much use of them without first finding a more practical expression for $\boldsymbol{M}_{\mathrm{CM}}$.

Since $\boldsymbol{r}$ is fixed in the rest frame, $\dot{\boldsymbol{r}} = \boldsymbol{\omega} \wedge \boldsymbol{r}$. Therefore

$$\begin{aligned} \boldsymbol{M}_{\mathrm{CM}} &= \int \rho \boldsymbol{r} \wedge (\boldsymbol{\omega} \wedge \boldsymbol{r})\, \mathrm{d}\tau \\ &= \int \rho[(\boldsymbol{r}\,.\,\boldsymbol{r})\boldsymbol{\omega} - (\boldsymbol{r}\,.\,\boldsymbol{\omega})\boldsymbol{r}]\, \mathrm{d}\tau. \end{aligned} \tag{3.2.4}$$

On taking components in $(\boldsymbol{e}_1, \boldsymbol{e}_2, \boldsymbol{e}_3)$ (the orthonormal triad fixed in the body), this becomes

$$\begin{aligned} M_i &= \int \rho(r_k r_k \omega_i - r_j \omega_j r_i)\, \mathrm{d}\tau \\ &= \omega_j \int \rho(r_k r_k \delta_{ij} - r_i r_j)\, \mathrm{d}\tau \\ &= J_{ij}\omega_j \end{aligned} \tag{3.2.5}$$

where M_i and ω_i are the $(\boldsymbol{e}_1, \boldsymbol{e}_2, \boldsymbol{e}_3)$-components of $\boldsymbol{M}_{\mathrm{CM}}$ and $\boldsymbol{\omega}$; and the J_{ij} are the entries in the inertia matrix $J(R)$. We can write down the components of $\boldsymbol{M}_{\mathrm{CM}}$, therefore, once we know the angular velocity $\boldsymbol{\omega}$ and the inertia matrix at the centre of mass. Note that $\boldsymbol{M}_{\mathrm{CM}}$ is parallel to $\boldsymbol{\omega}$ if and only if $\boldsymbol{\omega}$ is aligned with a principal axis.

Now take the axes of R to be along principal axes at O. Then eqn (3.2.5) reduces to

$$M_1 = A\omega_1, \quad M_2 = B\omega_2, \quad M_3 = C\omega_3 \tag{3.2.6}$$

where A, B, and C are the principal moments of inertia at O. It follows that $\mathrm{D}\boldsymbol{M}_{\mathrm{CM}}$, which is the time derivative of $\boldsymbol{M}_{\mathrm{CM}}$ with respect to R, has components $A\dot{\omega}_1$, $B\dot{\omega}_2$, and $C\dot{\omega}_3$. By the Coriolis theorem, however,

$$\dot{\boldsymbol{M}}_{\mathrm{CM}} = \mathrm{D}\boldsymbol{M}_{\mathrm{CM}} + \boldsymbol{\omega} \wedge \boldsymbol{M}_{\mathrm{CM}}. \tag{3.2.7}$$

where the dot again denotes the time derivative with respect to $\hat{R}$. Therefore

$$\begin{aligned} A\dot{\omega}_1 + (C - B)\omega_2\omega_3 &= G_1 \\ B\dot{\omega}_2 + (A - C)\omega_3\omega_1 &= G_2 \\ C\dot{\omega}_3 + (B - A)\omega_1\omega_2 &= G_3 \end{aligned} \tag{3.2.8}$$

where G_1, G_2, and G_3 are the components of $\sum \boldsymbol{r} \wedge \boldsymbol{E}$ in the rest frame R. These are *Euler's equations*. They determine the time-dependence of the angular velocity and hence of the orientation of the body.

We can also take moments about the origin of the inertial frame, to obtain

$$\dot{\boldsymbol{M}}_{\hat{O}} = \sum (\boldsymbol{r} + \boldsymbol{x}) \wedge \boldsymbol{E}, \tag{3.2.9}$$

where $\boldsymbol{M}_{\hat{O}}$ is the angular momentum about $\hat{O}$:

$$\boldsymbol{M}_{\hat{O}} = \boldsymbol{M}_{\text{CM}} + \boldsymbol{x} \wedge \boldsymbol{p}. \tag{3.2.10}$$

The right-hand side of eqn (3.2.9) is the moment of the external forces about $\hat{O}$.

Equation (3.2.9) does not contain any new information: it is an immediate consequence of eqns (3.2.3). It is particularly useful, however, in the case of a body rotating about a fixed point; in other words, when there is a point P of the body which is at rest relative to the inertial frame at all times. We can then translate the origin of both the inertial frame and the rest frame of the body to P, so that now $R = (P, (\boldsymbol{e}_1, \boldsymbol{e}_2, \boldsymbol{e}_3))$ and $\hat{R} = (P, (\hat{\boldsymbol{e}}_1, \hat{\boldsymbol{e}}_2, \hat{\boldsymbol{e}}_3))$.

Since P is fixed in both frames, the velocity relative to $\hat{R}$ of the point of the body with P-position vector $\boldsymbol{r}$ is $\boldsymbol{\omega} \wedge \boldsymbol{r}$. Therefore

$$\boldsymbol{M}_P = \int \rho \boldsymbol{r} \wedge (\boldsymbol{\omega} \wedge \boldsymbol{r})\, \mathrm{d}\tau. \tag{3.2.11}$$

Hence the $(\boldsymbol{e}_1, \boldsymbol{e}_2, \boldsymbol{e}_3)$-components M_1, M_2, M_3 of $\boldsymbol{M}_P$ are again given by eqn (3.2.5), only now the J_{ij} are the entries in the inertia matrix at P; and when the axes of R are aligned with principal axes at P, then we once again obtain Euler's equations for the $(\boldsymbol{e}_1, \boldsymbol{e}_2, \boldsymbol{e}_3)$-components of $\boldsymbol{\omega}$, only now A, B, and C are the principal moments of inertia at P and G_1, G_2, and G_3 are the $(\boldsymbol{e}_1, \boldsymbol{e}_2, \boldsymbol{e}_3)$-components of the total moment of the external forces about P.

It is very important to remember that the expression (3.2.5) for the components of the angular momentum is valid *only* for

1) angular momentum about the centre of mass in the case of arbitrary motion; or
2) in the case of rotation about a fixed point of an inertial frame, for the angular momentum about the fixed point.

Similarly, Euler's equations for the components of $\boldsymbol{\omega}$ hold only if the rest frame R either has its origin at the centre of mass (general motion) or, in the case of rotation about a fixed point of $\hat{R}$, has its origin at the fixed point.

Example (3.2.1) Free rotation. Consider a rigid body rotating about a fixed point P. Let $R = (P, (\boldsymbol{e}_1, \boldsymbol{e}_2, \boldsymbol{e}_3))$ be a rest frame with its axes aligned with the principal axes at P and suppose that $G_1 = G_2 = G_3 = 0$. Then Euler's equations reduce to

$$\begin{aligned} A\dot{\omega}_1 + (C - B)\omega_2\omega_3 &= 0 \\ B\dot{\omega}_2 + (A - C)\omega_3\omega_1 &= 0 \\ C\dot{\omega}_3 + (B - A)\omega_1\omega_2 &= 0. \end{aligned} \tag{3.2.12}$$

We shall assume that $A < B < C$.

By multiplying the three equations first by ω_1, ω_2, and ω_3 respectively and adding; and second by $A\omega_1$, $B\omega_2$, $C\omega_3$ and adding, we find that

$$A\omega_1^2 + B\omega_2^2 + C\omega_3^2 = 2T \qquad (3.2.13)$$

$$A^2\omega_1^2 + B^2\omega_2^2 + C^2\omega_3^2 = M^2 \qquad (3.2.14)$$

where T and M are constants; T is the kinetic energy and M is the magnitude of the angular momentum vector $\boldsymbol{M}_P$.

The instantaneous axis is the line through P in the direction of $\boldsymbol{\omega}$. By studying the way in which it moves relative to R, we can discover a great deal about the dynamical behaviour of the body.

By writing $\boldsymbol{\omega} = \lambda \boldsymbol{r}$ and eliminating λ between eqns (3.2.13) and (3.2.14), we see that the instantaneous axis is a generator of a fixed quadric cone in the body. In the rest frame R, the cone has equation

$$\alpha x^2 + \beta y^2 + \gamma z^2 = 0 \qquad (3.2.15)$$

where x, y, and z are the $(\boldsymbol{e}_1, \boldsymbol{e}_2, \boldsymbol{e}_3)$-components of $\boldsymbol{r}$ and

$$\alpha = \frac{A}{2T} - \frac{A^2}{M^2}$$

$$\beta = \frac{B}{2T} - \frac{B^2}{M^2} \qquad (3.2.16)$$

$$\gamma = \frac{C}{2T} - \frac{C^2}{M^2}.$$

With A, B, and C fixed, the shape of the cone depends on the value of $M^2/2T$ (which must lie between A and C). There are three cases.

(1) $M^2 = 2BT$. Then $\beta = 0$, $\alpha > 0$, and $\gamma < 0$. The cone degenerates into a pair of planes that intersect along the y-axis.

(2) $2CT \geqslant M^2 > 2BT$. Then $\alpha > 0$, $\beta > 0$, and $\gamma \leqslant 0$. The intersection of the cone with the plane $z = 1$ is one of the family of ellipses in Fig. 3.2.1.

(3) $2BT > M^2 \geqslant 2AT$. Then $\alpha \geqslant 0$, $\beta < 0$, and $\gamma < 0$. The intersection of the cone with the plane $x = 1$ is one of the family of ellipses in Fig. 3.2.2.

In (2) and (3), the limiting case of the parallel lines arises as $\beta \to 0$. The arrows on the ellipses indicate the movement of the instantaneous axis relative to the body: the directions are found by examining the signs in Euler's equations.

By combining figs. 3.2.1 and 3.2.2, we obtain Fig. 3.2.3, which shows the quadric cone and the direction of movement of the instantaneous axis for various values of $M^2/2T$. (In the limiting case $A = B$, the plane pair coalesces into the xy-plane and the quadric cones become circular: the picture reduces to Fig. 3.2.4).

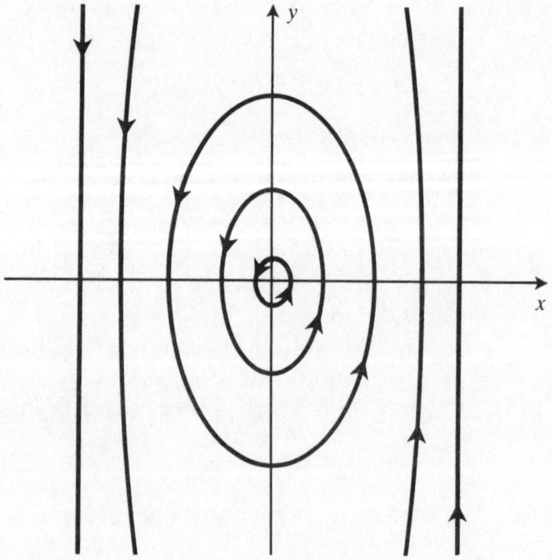

Fig. 3.2.1 Case (2): the intersection with the plane $z = 1$ is one of the ellipses $\alpha x^2 + \beta y^2 = -\gamma$.

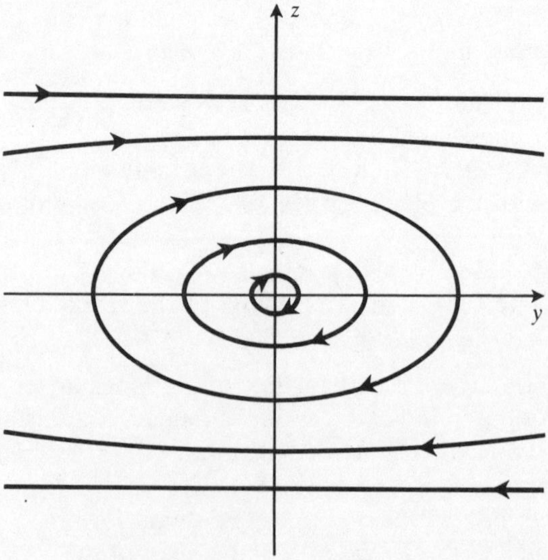

Fig. 3.2.2 Case (3): The intersection with the plane $x = 1$ is one of the ellipses $-\beta y^2 - \gamma z^2 = \alpha$.

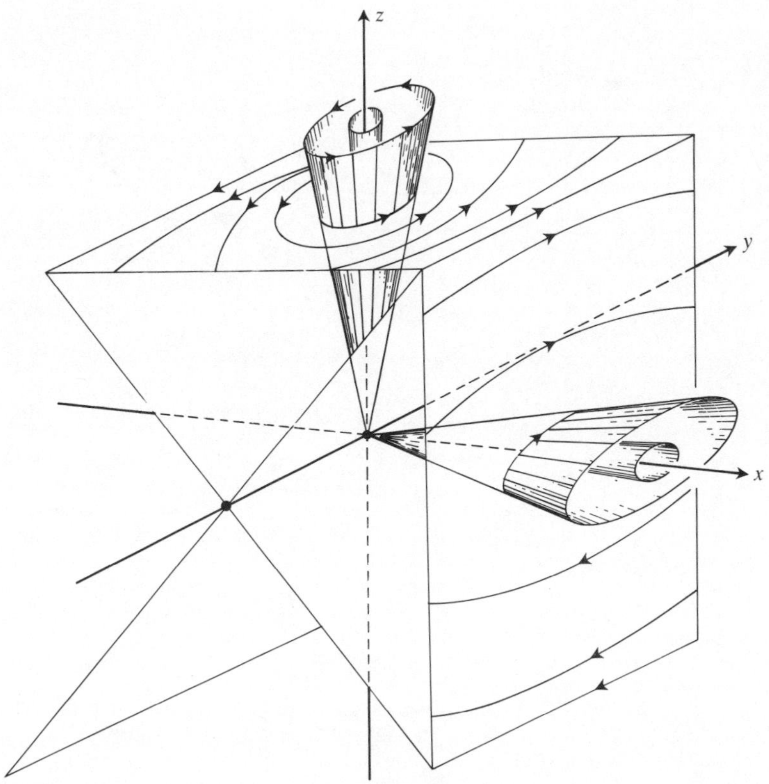

Fig. 3.2.3

Motion with constant angular velocity is possible about any of the three coordinate axes of the rest frame: $\boldsymbol{\omega}$ is then parallel to $\boldsymbol{M}_P$ and the instantaneous axis is fixed both in the body and in the inertial frame. The motion is stable if $\omega_2 = \omega_3 = 0$ or if $\omega_1 = \omega_2 = 0$: if the body is disturbed slightly when it is rotating about either the x-axis or the z-axis, then the instantaneous axis must remain on a narrow cone surrounding the original axis of rotation. However, the motion is unstable when $\omega_1 = \omega_3 = 0$.

The case $M^2 = 2BT$ is worth investigating further. By substituting for ω_1 and ω_3 from eqns (3.2.13) and (3.2.14) into the second of eqns (3.2.12),

$$B^2\dot{\omega}_2^2 = \frac{(2T - B\omega_2^2)^2(C-B)(B-A)}{AC}. \tag{3.2.17}$$

This can be solved analytically; but it is possible to deduce the qualitative behaviour by sketching the orbits in the ω_2, $\dot{\omega}_2$-plane (Fig. 3.2.5) without

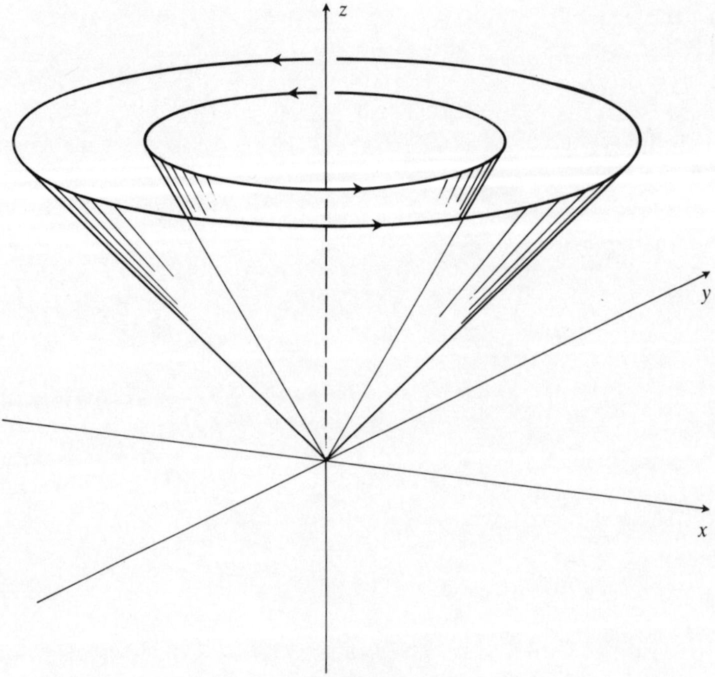

Fig. 3.2.4

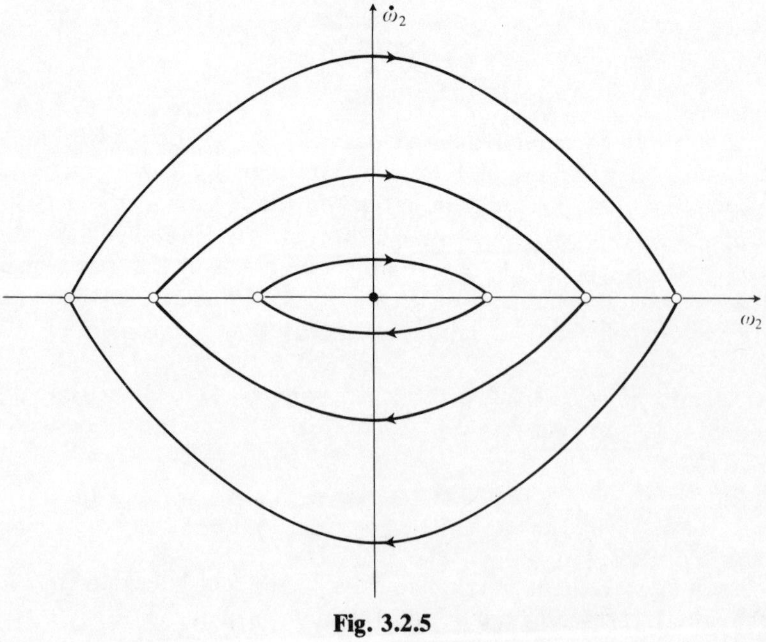

Fig. 3.2.5

finding the explicit solutions. The orbits are the curves given by eqn (3.2.17) for the different values of T. Unstable equilibrium is possible when $\omega_2 = \pm\sqrt{(2T/B)}$ (rotation about the y-axis). But if the motion is disturbed (with the relationship between T and M^2 maintained), then the system follows one of the curves in Fig. 3.2.5 and ω_2 tends asymptotically to the negative of its original value. The instantaneous axis sweeps out one or other of the $M^2 = 2BT$ pair of planes in the body, eventually coming back into coincidence with the y-axis, but with ω_2 having changed sign. During the process the angular momentum vector remains fixed relative to the inertial frame. Thus the effect of the disturbance is to cause the body to flip over: if the intersection points of the y-axis of R with the surface are marked red and blue, then, seen from the inertial frame, the body ends up spinning in exactly the same way as initially, but with the red and blue points interchanged.

This somewhat implausible behaviour depends, of course, on the exact relationship between T and M^2 being maintained when the original rotation about the y-axis is disturbed.

Example (3.2.2) Motion of a top: vector treatment. A top is a rigid body with axial symmetry about a line through its centre of mass.

Suppose that P is a point on the axis of symmetry at a distance a from the centre of mass. Consider the motion in which P is fixed relative to an inertial frame and the top is rotating about P without friction under the influence of gravity.

Let $\boldsymbol{e}$ be the unit vector along the axis of symmetry pointing from P towards the centre of mass (Fig. 3.2.6). Then

$$\dot{\boldsymbol{e}} = \boldsymbol{\omega} \wedge \boldsymbol{e} \tag{3.2.18}$$

where $\boldsymbol{\omega}$ is the angular velocity of the top and the dot denotes the time derivative with respect to the inertial frame. By taking the vector product with $\boldsymbol{e}$,

$$\boldsymbol{\omega} = \boldsymbol{e} \wedge \dot{\boldsymbol{e}} + n\boldsymbol{e} \tag{3.2.19}$$

where $n = \boldsymbol{\omega} \cdot \boldsymbol{e}$.

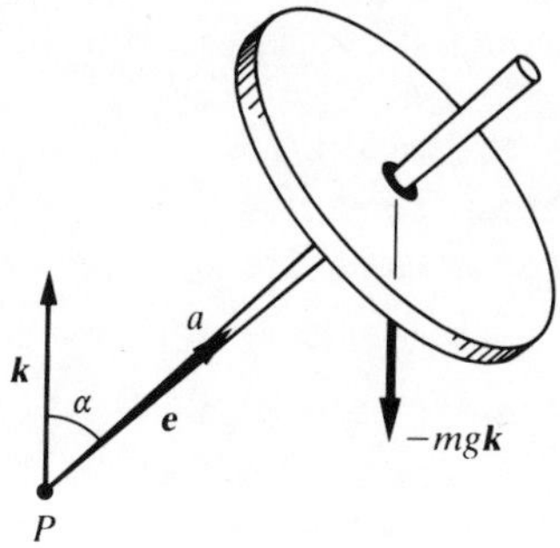

Fig. 3.2.6

The principal axes at P must be the axis of symmetry itself and all the lines through P orthogonal to $\boldsymbol{e}$. Let the corresponding principal moments of inertia be C (axis of symmetry) and A (any orthogonal axis).

Since $\boldsymbol{e}$ is along the axis of symmetry and $\boldsymbol{e} \wedge \dot{\boldsymbol{e}}$ is orthogonal to the axis of symmetry,

$$\boldsymbol{M}_P = A\boldsymbol{e} \wedge \dot{\boldsymbol{e}} + Cn\boldsymbol{e}. \tag{3.2.20}$$

The only force with a nonzero moment about P is gravity, which acts through the centre of mass. Therefore

$$\dot{\boldsymbol{M}}_P = A\boldsymbol{e} \wedge \ddot{\boldsymbol{e}} + C\dot{n}\boldsymbol{e} + Cn\dot{\boldsymbol{e}} = -mga\boldsymbol{e} \wedge \boldsymbol{k} \tag{3.2.21}$$

where $\boldsymbol{k}$ is the unit vector in the direction of the upward vertical.

By taking the scalar product with $\boldsymbol{e}$, we deduce that $\dot{n} = 0$.

We shall not investigate the motion in more detail at this stage, except to remark that eqn (3.2.21) has solutions for which

$$\dot{\boldsymbol{e}} = \Omega \boldsymbol{k} \wedge \boldsymbol{e}, \tag{3.2.22}$$

where Ω is constant. These correspond to *steady precession,* in which $\boldsymbol{e} \,.\, \boldsymbol{k}$ is constant. The axis of symmetry makes a fixed angle with the vertical and rotates about the vertical with constant angular speed Ω.

On substituting for $\dot{\boldsymbol{e}}$ from eqn (3.2.22) into eqn (3.2.21), one finds that in steady precession, n, Ω, and the angle α between $\boldsymbol{e}$ and $\boldsymbol{k}$ are related by

$$A\Omega^2 \cos \alpha - Cn\Omega + mga = 0. \tag{3.3.23}$$

Exercises

(3.2.1) Solve eqn (3.2.17) and verify the statements made about the behaviour of ω_2.

(3.2.2) In example (3.2.1) sketch and interpret the orbits of the system in the ω_2, $\dot{\omega}_2$-plane for different values of M^2 and T.

(3.2.3) The surface of a rigid body is an ellipsoid with equation $Ax^2 + By^2 + Cz^2 = k^2$ in a rest frame with origin O at the centre of mass and axes aligned with the principal axes at the centre of mass; A, B, and C are the principal moments of inertia at O and k is a constant.

Show that if the body rotates freely about O, with O fixed relative to an inertial frame, then the two tangent planes to the surface at the intersection points between the surface and the instantaneous axis are fixed relative to the inertial frame. Deduce that the body moves as if it were rolling between these two planes.

(3.2.4)† In example (3.2.2): let $\boldsymbol{N} = \boldsymbol{e} \wedge \dot{\boldsymbol{e}}$ and let θ be the angle between $\boldsymbol{e}$ and $\boldsymbol{k}$. Show that

$$A\boldsymbol{N} \,.\, \boldsymbol{N} + 2mga \cos\theta \quad \text{and} \quad A\boldsymbol{N} \,.\, \boldsymbol{k} + Cn \cos\theta$$

are constant.

Show that if initially $\theta = \beta$ and $\boldsymbol{N} = 0$, then during the subsequent motion

$$2Amga/C^2n^2 \geqslant \cos\beta - \cos\theta \geqslant 0.$$

(3.2.5)† A uniform solid sphere of radius a rolls without slipping inside a fixed sphere of radius $2a$. Show that if $\boldsymbol{e}$ is the unit vector pointing from the centre of the larger sphere towards the centre of the smaller sphere, then

$$7a\boldsymbol{e} \wedge \ddot{\boldsymbol{e}} - 2an\dot{\boldsymbol{e}} + 5g\boldsymbol{e} \wedge \boldsymbol{k} = 0$$

where n is constant and $\boldsymbol{k}$ is a unit vector in the direction of the upward vertical. The dot denotes the time derivative with respect to fixed axes.

(3.2.6)† A hollow right circular cylinder is fixed with its axis vertical. A uniform solid sphere rolls without slipping on the inside surface of the cylinder. Show that when the centre of the sphere does not move on a vertical line, the height of the centre of the sphere performs simple harmonic motion, and that between oscillations the plane containing the axis of the cylinder and the centre of the sphere turns through an angle $\pi\sqrt{14}$.

3.3 Lagrange's equations

Euler's equations determine the time-dependence of the angular velocity $\boldsymbol{\omega}$ of a rigid body; but, except in special cases, they are not a good starting point for determining the evolution of the configuration. For that, we need coordinates in the configuration space and the corresponding Lagrangian equations.

There will be six coordinates in all: for example, the three Cartesian coordinates of the centre of mass (in some inertial frame) and three angular coordinates for the rotational degrees of freedom.

It is the angular coordinates that we shall now define. Let $B = (\boldsymbol{e}_1, \boldsymbol{e}_2, \boldsymbol{e}_3)$ and $\hat{B} = (\hat{\boldsymbol{e}}_1, \hat{\boldsymbol{e}}_2, \hat{\boldsymbol{e}}_3)$ be two orthonormal triads and let H be the transition matrix from $\hat{B}$ to B, so that $H_{ij} = \boldsymbol{e}_i \cdot \hat{\boldsymbol{e}}_j$. The nine H_{ij} determine the relative orientation of the two triads. But we cannot specify them independently of each other, since they must satisfy the orthogonality relations

$$H_{ik}H_{jk} = \delta_{ij}. \tag{3.3.1}$$

There are six independent equations here (for example $i = j = 1$; $i = j = 2$; $i = j = 3$; $i = 1$ and $j = 2$; $i = 2$ and $j = 3$; $i = 3$ and $j = 1$), so we should be able to express the H_{ij} in terms of three independent parameters. The following proposition shows that this is indeed the case.

Proposition (3.3.1). Let $B=(\boldsymbol{e}_1, \boldsymbol{e}_2, \boldsymbol{e}_3)$ and $\hat{B}=(\hat{\boldsymbol{e}}_1, \hat{\boldsymbol{e}}_2, \hat{\boldsymbol{e}}_3)$ be two orthonormal triads. Then the transition matrix H from $\hat{B}$ to B can be written as a product

$$H=\begin{pmatrix} \cos\psi & \sin\psi & 0 \\ -\sin\psi & \cos\psi & 0 \\ 0 & 0 & 1 \end{pmatrix}\begin{pmatrix} \cos\theta & 0 & -\sin\theta \\ 0 & 1 & 0 \\ \sin\theta & 0 & \cos\theta \end{pmatrix}\begin{pmatrix} \cos\varphi & \sin\varphi & 0 \\ -\sin\varphi & \cos\varphi & 0 \\ 0 & 0 & 1 \end{pmatrix} \tag{3.3.2}$$

where $0\leqslant\psi<2\pi$, $0\leqslant\theta\leqslant\pi$, $0\leqslant\varphi<2\pi$. Moreover ψ, θ, and φ are uniquely determined by H provided that $|H_{33}|\neq 1$.

Proof. First uniqueness: if ψ, θ, and φ exist, then

$$\begin{aligned} &H_{33}=\cos\theta, \\ &H_{31}=\sin\theta\cos\varphi, \qquad H_{32}=\sin\theta\sin\varphi, \\ &H_{13}=-\sin\theta\cos\psi, \qquad H_{23}=\sin\theta\sin\psi. \end{aligned} \tag{3.3.3}$$

The first equation fixes the value of θ uniquely in the interval $[0,\pi]$; the second pair then determine φ uniquely in $[0,2\pi)$, provided that $\sin\theta\neq 0$; that is, provided that $|H_{33}|\neq 1$. Finally, under the same condition, the last pair determine ψ uniquely in $[0,2\pi)$.

To establish existence, we shall introduce two intermediate triads, $B'=(\boldsymbol{e}_1', \boldsymbol{e}_2', \boldsymbol{e}_3')$ and $B''=(\boldsymbol{e}_1'', \boldsymbol{e}_2'', \boldsymbol{e}_3'')$.

Suppose that $|H_{33}|=|\boldsymbol{e}_3\cdot\hat{\boldsymbol{e}}_3|\neq 1$. Then $\hat{\boldsymbol{e}}_3\wedge\boldsymbol{e}_3\neq 0$.

Let $\boldsymbol{j}$ be the unit vector in the direction of $\hat{\boldsymbol{e}}_3\wedge\boldsymbol{e}_3$. If $\theta\in(0,\pi)$ is the angle between $\boldsymbol{e}_3$ and $\hat{\boldsymbol{e}}_3$, then $\boldsymbol{e}_3\cdot\hat{\boldsymbol{e}}_3=\cos\theta$ and $\hat{\boldsymbol{e}}_3\wedge\boldsymbol{e}_3=\sin\theta\,\boldsymbol{j}$.

Since $\boldsymbol{j}$ is a unit vector orthogonal to $\boldsymbol{e}_3$,

$$\boldsymbol{j}=\sin\psi\,\boldsymbol{e}_1+\cos\psi\,\boldsymbol{e}_2 \tag{3.3.4}$$

for some $\psi\in[0,2\pi)$; and since $\boldsymbol{j}$ is also orthogonal to $\hat{\boldsymbol{e}}_3$,

$$\boldsymbol{j}=-\sin\varphi\,\hat{\boldsymbol{e}}_1+\cos\varphi\,\hat{\boldsymbol{e}}_2 \tag{3.3.5}$$

for some $\varphi\in[0,2\pi)$ (Fig. 3.3.1).

Define B' by

$$\boldsymbol{e}_1'=\boldsymbol{j}\wedge\boldsymbol{e}_3, \qquad \boldsymbol{e}_2'=\boldsymbol{j}, \qquad \boldsymbol{e}_3'=\boldsymbol{e}_3 \tag{3.3.6}$$

and let K be the transition matrix from B' to B, so that $K_{ij}=\boldsymbol{e}_i\cdot\boldsymbol{e}_j'$. Then, from eqn (3.3.4),

$$K=\begin{pmatrix} \cos\psi & \sin\psi & 0 \\ -\sin\psi & \cos\psi & 0 \\ 0 & 0 & 1 \end{pmatrix}. \tag{3.3.7}$$

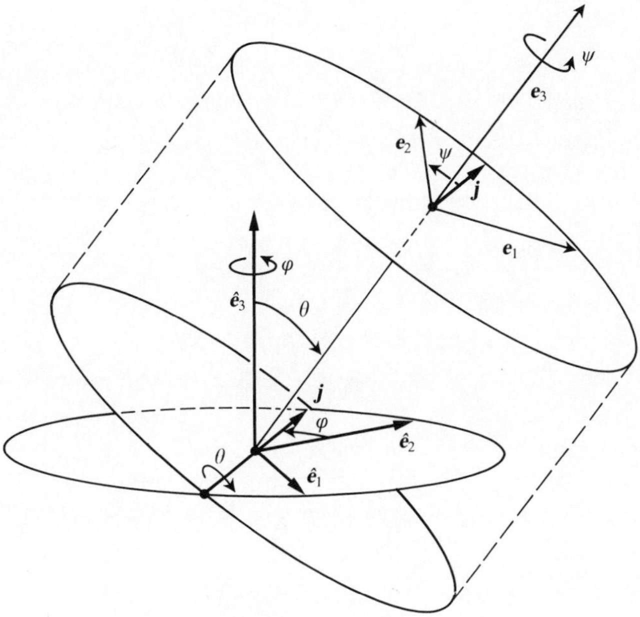

Fig. 3.3.1 Euler angles.

Similarly, define B'' by

$$\boldsymbol{e}_1'' = \boldsymbol{j} \wedge \hat{\boldsymbol{e}}_3, \quad \boldsymbol{e}_2'' = \boldsymbol{j}, \quad \boldsymbol{e}_3'' = \hat{\boldsymbol{e}}_3, \tag{3.3.8}$$

and let M be the transition matrix from $\hat{B}$ to B'', so that $M_{ij} = \boldsymbol{e}_i'' \cdot \hat{\boldsymbol{e}}_j$. Then, from eqn (3.3.5),

$$M = \begin{pmatrix} \cos\varphi & \sin\varphi & 0 \\ -\sin\varphi & \cos\varphi & 0 \\ 0 & 0 & 1 \end{pmatrix}. \tag{3.3.9}$$

Finally, let L be the transition matrix from B'' to B', so that $L_{ij} = \boldsymbol{e}_i' \cdot \boldsymbol{e}_j''$. Then, from eqns (3.3.6) and (3.3.8),

$$L = \begin{pmatrix} \cos\theta & 0 & -\sin\theta \\ 0 & 1 & 0 \\ \sin\theta & 0 & \cos\theta \end{pmatrix} \tag{3.3.10}$$

since, for example,

$$\begin{aligned} \boldsymbol{e}_1' \cdot \boldsymbol{e}_3'' &= (\boldsymbol{j} \wedge \boldsymbol{e}_3) \cdot \hat{\boldsymbol{e}}_3 \\ &= (\boldsymbol{e}_3 \wedge \hat{\boldsymbol{e}}_3) \cdot \boldsymbol{j} \\ &= -\sin\theta. \end{aligned} \tag{3.3.11}$$

Now $e'_q = K_{kq}e_k$, $e''_p = L_{qp}e'_q$, and $\hat{e}_j = M_{pj}e''_p$. By putting these together,

$$\hat{e}_j = H_{kj}e_k = K_{kq}L_{qp}M_{pj}e_k. \tag{3.3.12}$$

Hence $H = KLM$.

In the case $H_{33} = \pm 1$,

$$H = \begin{pmatrix} \pm\cos\alpha & \pm\sin\alpha & 0 \\ -\sin\alpha & \cos\alpha & 0 \\ 0 & 0 & \pm 1 \end{pmatrix} \tag{3.3.13}$$

for some $\alpha \in [0, 2\pi)$ (see exercise (1.2.13)). We can take $\psi = 0$, $\varphi = \alpha$, and $\theta = 0$ $(H_{33} = 1)$ or $\theta = \pi$ $(H_{33} = -1)$, although other choices are possible.

□

Definition (3.3.1). The angles θ, φ, and ψ are the *Euler angles* of B relative to $\hat{B}$.

Note from eqn (3.3.3) that e_3 has $\hat{B}$-components

$$(\sin\theta\cos\varphi, \sin\theta\sin\varphi, \cos\theta) \tag{3.3.14}$$

so that θ and φ are the spherical polar angles of e_3 in $\hat{B}$; and that $\hat{e}_3$ has B-components

$$(-\sin\theta\cos\psi, \sin\theta\sin\psi, \cos\theta) \tag{3.3.15}$$

so that θ and $\pi - \psi$ are the spherical polar angles of $\hat{e}_3$ in B.

The proof splits the transformation from $\hat{B}$ to B into three steps: a rotation about $\hat{e}_3$ through φ, which brings e_2 into coincidence with j; a rotation about j through θ, which brings $\hat{e}_3$ into coincidence with e_3; and, finally, a rotation about e_3 through ψ, which brings j into coincidence with e_2 (Fig. 3.3.1).

We can specify the changing orientation of B relative to $\hat{B}$ at any time by giving θ, φ, and ψ as functions of time; and we can express the angular velocity of B relative to $\hat{B}$ in terms of the derivatives of the Euler angles, by making use of proposition (1.2.3). The angular velocity of B relative to B' is $\dot{\psi}e_3$ (see example (1.2.2)); the angular velocity of B' relative to B'' is $\dot{\theta}j$; and the angular velocity of B'' relative to $\hat{B}$ is $\dot{\varphi}\hat{e}_3$. Therefore the angular velocity of B relative to $\hat{B}$ is

$$\begin{aligned} \omega &= \dot{\psi}e_3 + \dot{\theta}j + \dot{\varphi}\hat{e}_3 \\ &= \dot{\psi}e_3 + \dot{\theta}(\sin\psi\, e_1 + \cos\psi\, e_2) \\ &\quad + \dot{\varphi}(-\sin\theta\cos\psi\, e_1 + \sin\theta\sin\psi\, e_2 + \cos\theta\, e_3), \end{aligned} \tag{3.3.16}$$

where we have substituted for $\boldsymbol{j}$ and $\hat{\boldsymbol{e}}_3$ from eqns (3.3.4) and (3.3.15). Hence the B-components of $\boldsymbol{\omega}$ are

$$\begin{aligned}\omega_1 &= \dot{\theta}\sin\psi - \dot{\varphi}\sin\theta\cos\psi\\ \omega_2 &= \dot{\theta}\cos\psi + \dot{\varphi}\sin\theta\sin\psi\\ \omega_3 &= \dot{\psi} + \dot{\varphi}\cos\theta.\end{aligned} \tag{3.3.17}$$

Similarly, the $\hat{B}$-components are

$$\begin{aligned}\hat{\omega}_1 &= -\dot{\theta}\sin\varphi + \dot{\psi}\sin\theta\cos\varphi\\ \hat{\omega}_2 &= \dot{\theta}\cos\varphi + \dot{\psi}\sin\theta\sin\varphi\\ \hat{\omega}_3 &= \dot{\varphi} + \dot{\psi}\cos\theta\end{aligned} \tag{3.3.18}$$

(by using eqns (3.3.5) and (3.3.14)).

Put in less formal terms, $\boldsymbol{\omega}$ is the sum of three vectors: the first, $\dot{\psi}\boldsymbol{e}_3$, is the angular velocity of B relative to $\hat{B}$ when θ and φ are held constant; the second, $\dot{\theta}\boldsymbol{j}$, is the angular velocity when ψ and φ are held constant; and the third, $\dot{\varphi}\hat{\boldsymbol{e}}_3$, is the angular velocity when θ and ψ are held constant. This is the simplest way to obtain the angular velocity in practice.

Let $\hat{R} = (\hat{O}, (\hat{\boldsymbol{e}}_1, \hat{\boldsymbol{e}}_2, \hat{\boldsymbol{e}}_3))$ be an inertial frame and let $R = (O, (\boldsymbol{e}_1, \boldsymbol{e}_2, \boldsymbol{e}_3))$ be a rest frame of a rigid body. Suppose that O is at the centre of mass of the body and that the axes of R are aligned with the principal axes at O. Let x, y, and z be the coordinates of O in $\hat{R}$; and let θ, φ, and ψ be the Euler angles of $(\boldsymbol{e}_1, \boldsymbol{e}_2, \boldsymbol{e}_3)$ relative to $(\hat{\boldsymbol{e}}_1, \hat{\boldsymbol{e}}_2, \hat{\boldsymbol{e}}_3)$. Then x, y, z, θ, φ, and ψ are coordinates on the configuration space of the body. By combining eqn (3.1.29) with eqn (3.3.17), the kinetic energy of the body (relative to $\hat{R}$) is

$$\begin{aligned}T = {}&\tfrac{1}{2}m(\dot{x}^2 + \dot{y}^2 + \dot{z}^2) + \tfrac{1}{2}A(\dot{\theta}\sin\psi - \dot{\varphi}\sin\theta\cos\psi)^2\\ &+ \tfrac{1}{2}B(\dot{\theta}\cos\psi + \dot{\varphi}\sin\theta\sin\psi)^2 + \tfrac{1}{2}C(\dot{\psi} + \dot{\varphi}\cos\theta)^2,\end{aligned} \tag{3.3.19}$$

where A, B, and C are the principal moments of inertia at O.

Similarly, for rotation about a point P, which is fixed both in $\hat{R}$ and in the body, we can take $R = (P, (\boldsymbol{e}_1, \boldsymbol{e}_2, \boldsymbol{e}_3))$, with the $\boldsymbol{e}_i$ along principal axes at P. Then

$$\begin{aligned}T = {}&\tfrac{1}{2}A(\dot{\theta}\sin\psi - \dot{\varphi}\sin\theta\cos\psi)^2 + \tfrac{1}{2}B(\dot{\theta}\cos\psi + \dot{\varphi}\sin\theta\sin\psi)^2\\ &+ \tfrac{1}{2}C(\dot{\psi} + \dot{\varphi}\cos\theta)^2\end{aligned} \tag{3.3.20}$$

where A, B, and C are now the principal moments of inertia at P.

The coordinates are singular when $\theta = 0$, just as spherical polar coordinates are singular on the polar axis.

Example (3.3.1) Motion of a top: Lagrange's equations. We shall now look in more detail at the motion of the top described in example (3.2.2).

Choose $B = (\boldsymbol{e}_1, \boldsymbol{e}_2, \boldsymbol{e}_3)$ so that the $\boldsymbol{e}_i$ are aligned with the principal axes at P, with $\boldsymbol{e} = \boldsymbol{e}_3$ along the axis of symmetry; and choose $(\hat{\boldsymbol{e}}_1, \hat{\boldsymbol{e}}_2, \hat{\boldsymbol{e}}_3)$ so that $\boldsymbol{k} = \hat{\boldsymbol{e}}_3$. Then the Lagrangian is

$$L = \tfrac{1}{2}A(\dot{\varphi}^2 \sin^2\theta + \dot{\theta}^2) + \tfrac{1}{2}C(\dot{\psi} + \dot{\varphi}\cos\theta)^2 - mga\cos\theta, \quad (3.3.21)$$

by taking $A = B$ in eqn (3.3.20).

Since ψ and φ are cyclic

$$\begin{aligned} \dot{\psi} + \dot{\varphi}\cos\theta &= n \\ A\dot{\varphi}\sin^2\theta + Cn\cos\theta &= h \end{aligned} \quad (3.3.22)$$

where n and h are constants; and since $\partial L/\partial t = 0$,

$$A\dot{\theta}^2 + A\dot{\varphi}^2\sin^2\theta + Cn^2 + 2mga\cos\theta = 2E \quad (3.3.23)$$

where E is constant. The constants n, h, and E are, respectively, the $\boldsymbol{e}$-component of $\boldsymbol{\omega}$, the angular momentum about the vertical axis through P, and the total energy.

By writing $u = \cos\theta$ and rearranging the equations

$$\dot{\varphi} = \frac{h - Cnu}{A(1-u^2)} \quad \text{and} \quad A\dot{u}^2 = F(u) \quad (3.3.24)$$

where

$$F(u) = (2E - Cn^2 - 2mgau)(1-u^2) - \frac{(h - Cnu)^2}{A}. \quad (3.3.25)$$

The angles θ and φ are the spherical polar angles of $\boldsymbol{e}$: θ is the angle between the axis of the top and the upward vertical and $\dot{\varphi}$ is the angular speed with which the axis rotates about the vertical.

Suppose that the top is set in motion with $\theta = \cos^{-1}(u_1)$ and $\dot{\theta} = 0$. We shall keep h and n fixed (with $n > 0$ and $0 < h/Cn < 1$) and investigate what happens for various values of u_1 by looking at the phase portrait in the $u, \dot{u}$-plane. The orbits in the $u, \dot{u}$-plane are the curves $A\dot{u}^2 = F(u)$, the constant E in the definition of F being determined by

$$0 = (2E - Cn^2 - 2mgu_1)(1-u_1^2) - \frac{(h - Cnu_1)^2}{A}. \quad (3.3.26)$$

The graph of $F(u)$ for different values of u_1 and the phase portrait are shown in Figs. 3.3.2 and 3.3.3. The dotted vertical line is $u = h/Cn$: to the left of the line (that is for $-1 < u < h/Cn$), $\dot{\varphi}$ is positive; to the right $(h/Cn < u < 1)$, $\dot{\varphi}$ is negative. Remember that u must lie between -1 and 1.

As the top moves, its axis traces out a curve on the unit sphere centred at P. The rough form of the curve can be seen from the phase portrait (Fig. 3.3.3).

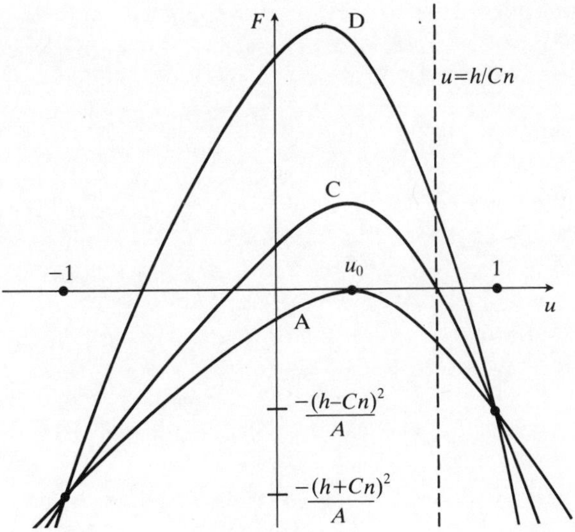

Fig. 3.3.2

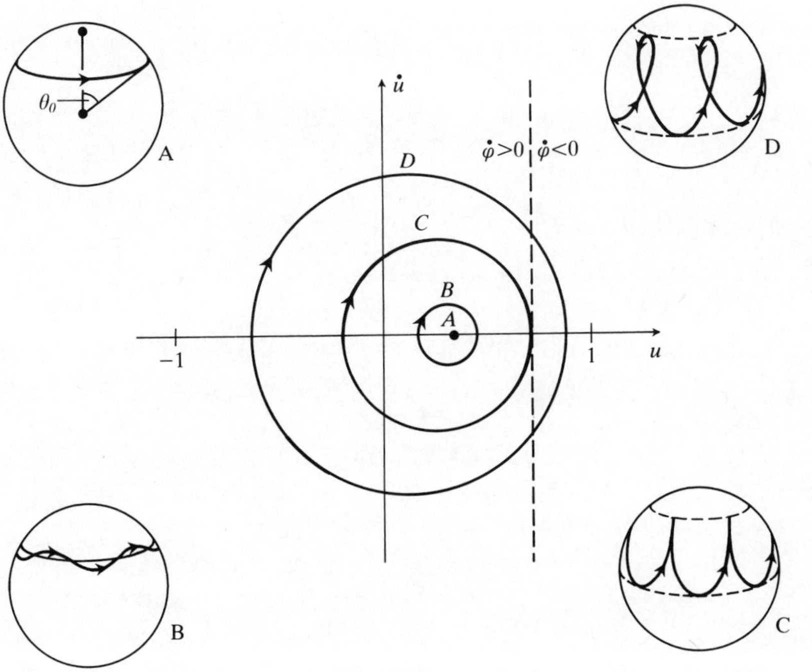

Fig. 3.3.3

There is a critical value u_0 of u_1 for which $u = u_1$ is a root of both $F(u)$ and $F'(u)$. Here the graph of $F(u)$ touches the u-axis at $u = u_0$ and the orbit is $u = u_0 = \text{constant}$. The axis precesses steadily about the vertical with θ and $\dot{\varphi}$ constant and $\dot{\varphi} > 0$. It traces out a horizontal circle on the sphere on which $\theta = \theta_0 = \cos^{-1}(u_0)$ (Fig. 3.3.3 A: top left).

For u_1 such that $u_0 < u_1 < h/Cn$, u oscillates between the two roots of $F(u)$ on either side of u_0, but $\dot{\varphi}$ is always positive: the axis rotates anticlockwise about the vertical (seen from above), but its angle with the vertical oscillates between the two corresponding values of θ (B: bottom left). The oscillatory behaviour of θ is called *nutation.*

If $u_1 = h/Cn$, then $\dot{\varphi} = 0$ when θ reaches its minimum (C: bottom right).

If $h/Cn < u_1 < 1$, then $\dot{\varphi}$ is negative during part of the motion and the curve on the sphere loops back on itself (D: top right).

Example (3.3.2).† A thin circular disc of radius a moves on a smooth horizontal plane. The disc makes contact with the plane at a point and can slip freely. Initially, the centre is at rest and the disc is spinning with angular speed n about its axis, which is at an angle α to the vertical.

The problem is to show that the spin about the axis (i.e. the component of the angular velocity along the axis) remains constant and that during the subsequent motion,

$$a\dot{\theta}^2(1 + 4\cos^2\theta) + 4an^2(\cos\alpha - \cos\theta)^2 \operatorname{cosec}^2\theta + 8g(\sin\theta - \sin\alpha) = 0, \tag{3.3.27}$$

where θ is the angle that the axis makes with the vertical.

Since there are no horizontal forces, the horizontal coordinates of the centre of mass remain constant: in effect, the system has three degrees of freedom.

Let $\hat{B} = (\hat{\boldsymbol{e}}_1, \hat{\boldsymbol{e}}_2, \hat{\boldsymbol{e}}_3)$ be a triad fixed relative to the plane, with $\hat{\boldsymbol{e}}_3$ normal to the plane; and let $B = (\boldsymbol{e}_1, \boldsymbol{e}_2, \boldsymbol{e}_3)$ be a triad fixed relative to the disc, with $\boldsymbol{e}_3$ normal to the disc.

We shall use the Euler angles of B relative to $\hat{B}$ as coordinates (Fig. 3.3.4).

The angular velocity of the disc is

$$\dot{\theta}\boldsymbol{j} + \dot{\psi}\boldsymbol{e}_3 + \dot{\varphi}\hat{\boldsymbol{e}}_3. \tag{3.3.28}$$

with $\boldsymbol{j}$ as in eqn (3.3.4). The height of the centre above the plane is $a\sin\theta$ and the velocity of the centre is $a\dot{\theta}\cos\theta\,\hat{\boldsymbol{e}}_3$.

The principal moments of inertia of the disc at its centre are $\frac{1}{4}ma^2$ about any diameter and $\frac{1}{2}ma^2$ about the axis through the centre normal to the plane of the disc. Therefore the Lagrangian is

$$L = \tfrac{1}{2}ma^2[\dot{\theta}^2\cos^2\theta + \tfrac{1}{4}(\dot{\theta}^2 + \dot{\varphi}^2\sin^2\theta) + \tfrac{1}{2}(\dot{\psi} + \dot{\varphi}\cos\theta)^2] - mga\sin\theta. \tag{3.3.29}$$

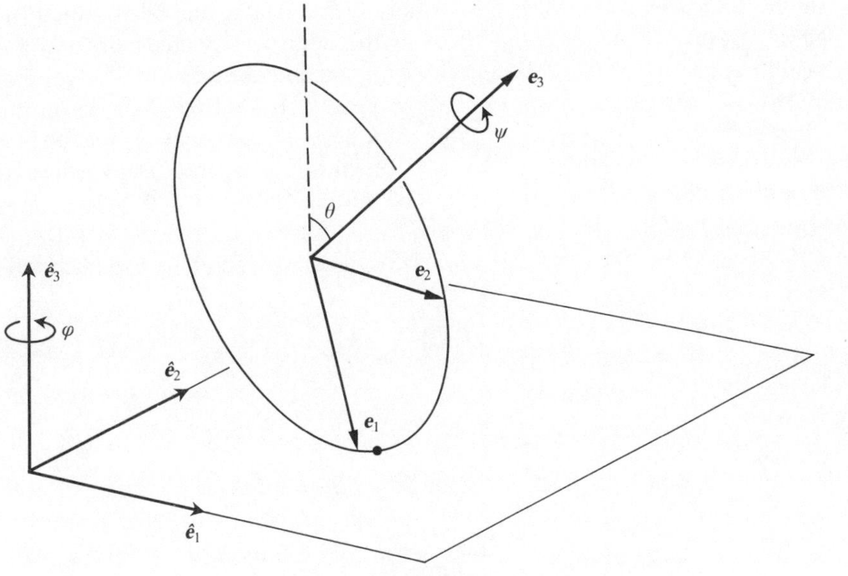

Fig. 3.3.4

Since ψ and φ are cyclic,

$$\begin{aligned} \dot{\psi} + \dot{\varphi} \cos \theta &= \text{constant} \\ \tfrac{1}{4}\dot{\varphi} \sin^2\theta + \tfrac{1}{2}(\dot{\psi} + \dot{\varphi} \cos \theta)\cos \theta &= \text{constant}. \end{aligned} \tag{3.3.30}$$

Now $\dot{\psi} + \dot{\varphi} \cos \theta = \boldsymbol{e}_3 \,.\, \boldsymbol{\omega}$ is the spin of the disc about its axis (note: a common mistake in this type of problem is to equate the spin about the axis to $\dot{\psi}$, forgetting the contribution from $\dot{\varphi}$). The first equation states that the spin is constant, as required.

Initially, $\theta = \alpha$, $\dot{\varphi} = \dot{\theta} = 0$, and $\boldsymbol{e}_3 \,.\, \boldsymbol{\omega} = n$. therefore

$$\dot{\psi} + \dot{\varphi} \cos \theta = n \quad \text{and} \quad \dot{\varphi} \sin^2\theta = 2n \cos \alpha - 2n \cos \theta \tag{3.3.31}$$

Since $\partial L/\partial t = 0$, the total energy

$$E = \tfrac{1}{2}ma^2[\tfrac{1}{4}\dot{\theta}^2(1 + 4\cos^2\theta) + (\cos \theta - \cos \alpha)^2 n^2 \operatorname{cosec}^2\theta + \tfrac{1}{2}n^2] + mga \sin \theta \tag{3.3.32}$$

is constant (proposition (2.5.2)). But initially $E = \tfrac{1}{4}ma^2n^2 + mga \sin \alpha$. Hence the stated result. □

Example (3.3.3).† Two uniform circular discs, each of radius a and mass m, are joined by a light rod AB of length $4a$ connected to the discs at their centres A, B in such a way that the planes of the discs are perpendicular to the rod and the discs can turn freely about the rod. The system is placed on a horizontal table which is rough enough to prevent slipping and which is forced to rotate with variable angular velocity Ω about a vertical axis passing through the centre O of AB. Initially the system is at rest. The problem is to find the angular velocity of the discs in terms of Ω.

Let $\boldsymbol{i}$ be a unit vector along OA, let $\boldsymbol{k}$ be a unit vertical vector, and let $\boldsymbol{j} = \boldsymbol{k} \wedge \boldsymbol{i}$. Let α be the angle between a fixed line on the disc with centre A and the vertical; and let φ be the angle between AB and a fixed horizontal line (see Fig. 3.3.5)

The velocity of A is $2a\dot{\varphi}\boldsymbol{j}$ and the angular velocity of the disc with centre A is

$$\boldsymbol{\omega} = \dot{\alpha}\boldsymbol{i} + \dot{\varphi}\boldsymbol{k}. \tag{3.3.33}$$

Hence its kinetic energy is

$$T = \tfrac{1}{2}m(2a\dot{\varphi})^2 + \tfrac{1}{2}(\tfrac{1}{2}ma^2\dot{\alpha}^2 + \tfrac{1}{4}ma^2\dot{\varphi}^2). \tag{3.3.34}$$

If the $\boldsymbol{j}$-component of the friction force on the disc is F, then

$$\begin{aligned} \frac{\mathrm{d}}{\mathrm{d}t}\left(\frac{\partial T}{\partial \dot{\varphi}}\right) &= \frac{\mathrm{d}}{\mathrm{d}t}(\tfrac{17}{4}ma^2\dot{\varphi}) = 2aF \\ \frac{\mathrm{d}}{\mathrm{d}t}\left(\frac{\partial T}{\partial \dot{\alpha}}\right) &= \frac{\mathrm{d}}{\mathrm{d}t}(\tfrac{1}{2}ma^2\dot{\alpha}) = aF \end{aligned} \tag{3.3.35}$$

(the generalized components of the friction force being found by

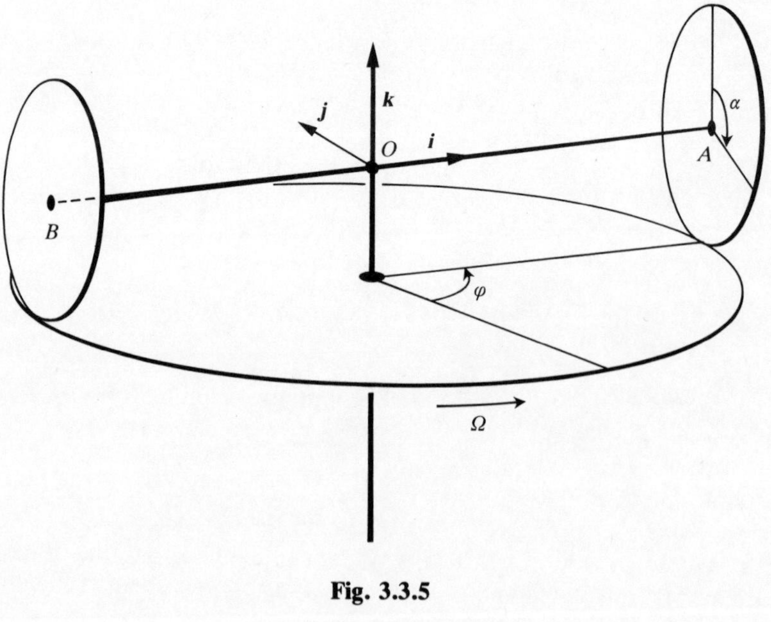

Fig. 3.3.5

considering the small displacements: (1) $\delta\varphi = \varepsilon$, $\delta\alpha = 0$ and (2) $\delta\varphi = 0$, $\delta\alpha = \varepsilon$). Therefore $17\dot{\varphi} = 4\dot{\alpha}$, since $\dot{\varphi}$ and $\dot{\alpha}$ are both zero initially.

Now the velocity of the point of the disc in contact with the table is

$$2a\dot{\varphi}\boldsymbol{j} + (\dot{\alpha}\boldsymbol{i} + \dot{\varphi}\boldsymbol{k}) \wedge (-a\boldsymbol{k}). \qquad (3.3.36)$$

This must be equal to the velocity of the point of the table in contact with the disc, which is $2a\Omega\boldsymbol{j}$. Therefore

$$2a\dot{\varphi} + a\dot{\alpha} = 2a\Omega, \qquad (3.3.37)$$

and so

$$\dot{\varphi} = \tfrac{8}{25}\Omega \quad \text{and} \quad \dot{\alpha} = \tfrac{34}{25}\Omega. \qquad (3.3.38)$$

□

Exercises

(3.3.1) In example (3.3.1):

(a) With $0 < h/Cn < 1$, show that u_0 is the root of

$$G(u) = (h - Cnu)(Cn - hu) - mgaA(1 - u^2)^2$$

in the interval $(-1, 1)$. By sketching the graph of G and considering $G(-1)$, $G(h/Cn)$ and $G(1)$, show that $u_0 < h/Cn$. Show that if h and n are large, then $u_0 \sim h/Cn$. Describe the motion of the axis of a top which is set in motion with large n, but with $\dot{\theta} = \dot{\varphi} = 0$.

(b) For small oscillations about steady precession with $\dot{\varphi} = \Omega$, show by considering $F''(u_0)$ that

$$A\ddot{v} + \lambda^2 v = 0$$

where $v = u - u_0$ and $\lambda^2 = A\Omega^2 - 2mgau_0 + m^2g^2a^2/A\Omega^2$. What is the period of the oscillations when n is large?

(c) Repeat the analysis in the example for the case $n > 0$, $-1 < h/Cn < 0$.

(3.3.2)† A thin uniform disc, mass M, radius a and centre C, has a thin uniform rod OC, mass m and length $a\sqrt{3}$, fixed to it at C, so that OC is orthogonal to the disc. The end O of the rod is fixed but freely pivoted at the centre O of a horizontal turntable, and the rim of the disc rests on the surface of the turntable. No slipping occurs.

The turntable is forced to rotate about the vertical axis through O with variable angular velocity Ω. Initially the system is at rest. Show that if P is the point of contact between the disc and the turntable, and φ is the angle between OP and a line fixed in the turntable, then

$$\dot{\varphi} = -\left(\frac{11M + 4m}{19M + 4m}\right)\Omega.$$

(3.3.3)† A thin uniform rod of length $2a$ and mass m has a small light ring fixed at one end. The ring is threaded on a fixed vertical wire. Show that if z is the height of the centre of the rod, θ the angle the rod makes with the upward vertical, and φ the angle that the vertical plane containing the rod makes with a fixed vertical plane, then the Lagrangian of the system is

$$L = \tfrac{1}{6}m[3\dot{z}^2 + a^2\dot{\theta}^2(1 + 3\cos^2\theta) + 4a^2\dot{\varphi}^2\sin^2\theta] - mgz.$$

Initially the rod makes an acute angle α with the vertical and its centre has velocity V perpendicular to the rod and the wire. Show that the angle the rod makes with the wire oscillates between α and $\pi - \alpha$ with period

$$\frac{a}{V}\int_{-\cos\alpha}^{\cos\alpha}\left[\frac{1 + 3u^2}{\cos^2\alpha - u^2}\right]^{1/2} du.$$

(3.3.4)† A uniform hollow circular cylinder of mass m, radius a, rolls without slipping on a fixed rough horizontal plane. A similar cylinder of mass m and the same length, but radius $\tfrac{1}{2}a$, rolls without slipping inside the larger cylinder. The two cylinders are positioned so that their axes are parallel and their ends coincide. Consider the vertical plane through the centre of mass. Show that if θ is the angle between the downward vertical and the line in this plane joining the centre of mass of the larger cylinder to a point fixed on the rim of the larger cylinder, and if φ is the angle between the downward vertical and the line joining the centres of mass, then

$$2ma^2\dot{\theta}^2 + \tfrac{1}{4}ma^2\dot{\varphi}^2 - \tfrac{1}{2}ma^2\dot{\theta}\dot{\varphi}(1 + \cos\varphi) - \tfrac{1}{2}mga\cos\varphi$$

is constant during the motion.

3.4 Nonholonomic constraints

We shall begin with an example.

Consider a uniform solid sphere of radius a and mass m, which is rolling without slipping on a moving horizontal plane. The plane passes through the origin of an inertial frame $\hat{R} = (\hat{O}, \hat{B})$ and is normal to $\hat{\mathbf{e}}_3$; the plane is being forced to move relative to $\hat{R}$ with velocity $\boldsymbol{U} = u\hat{\mathbf{e}}_1 + v\hat{\mathbf{e}}_2$, where u and v are given functions of time. We shall investigate the motion of the sphere, assuming that the only forces on it are the friction and normal reaction at the point of contact and the gravitational force $-mg\hat{\mathbf{e}}_3$.

Let $R = (O, B)$ be a frame fixed in the sphere, with its origin at the centre. Let $\boldsymbol{X} = x\hat{\mathbf{e}}_1 + y\hat{\mathbf{e}}_2 + a\hat{\mathbf{e}}_3$ be the vector from $\hat{O}$ to O and let θ, φ, and ψ be the Euler angles of B relative to $\hat{B}$ (Fig. 3.4.1).

All axes through the centre of the sphere are principal axes, with moment of inertia $\tfrac{2}{5}ma^2$. Therefore the sphere's total kinetic energy

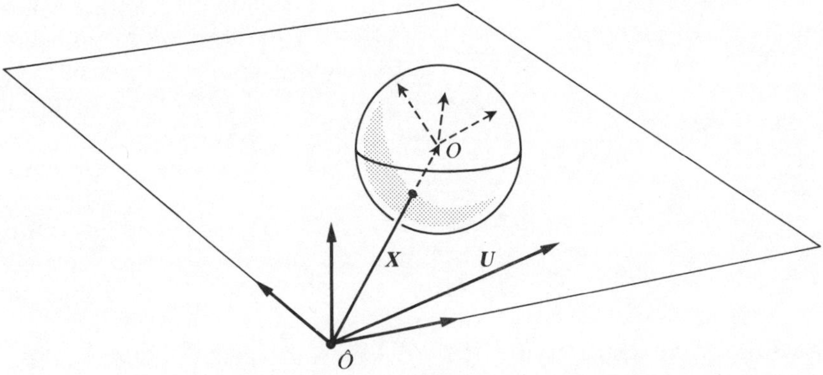

Fig. 3.4.1

relative to R is

$$T = \tfrac{1}{5}ma^2(\dot{\theta}^2 + \dot{\varphi}^2 + \dot{\psi}^2 + 2\dot{\varphi}\dot{\psi}\cos\theta) + \tfrac{1}{2}m(\dot{x}^2 + \dot{y}^2). \qquad (3.4.1)$$

If we eliminate the holonomic constraint that fixes the height of O above the plane, then we are left with five degrees of freedom, corresponding to the five generalized coordinates $q_1 = x$, $q_2 = y$, $q_3 = \theta$, $q_4 = \varphi$, and $q_5 = \psi$. The equations of motion are

$$\frac{\mathrm{d}}{\mathrm{d}t}\left(\frac{\partial T}{\partial v_a}\right) - \frac{\partial T}{\partial q_a} = K_a \qquad (3.4.2)$$

where the K_a are the q-components of the friction (gravity and the normal reaction do not contribute). The problem is to find the K_a, knowing only that they are the forces responsible for maintaining the rolling condition

$$\dot{\boldsymbol{X}} + \boldsymbol{\omega} \wedge (-a\hat{\boldsymbol{e}}_3) = \boldsymbol{U}, \qquad (3.4.3)$$

where $\boldsymbol{\omega}$ is the angular velocity of the sphere and the dot denotes the time derivative with respect to the inertial frame.

The $\hat{B}$-components of $\boldsymbol{\omega}$ are given by eqn (3.3.18); and by taking the $\hat{B}$-components of eqn (3.4.3), we obtain two constraint equations

$$\begin{aligned} \dot{x} - a\dot{\theta}\cos\varphi - a\dot{\psi}\sin\theta\sin\varphi &= u \\ \dot{y} - a\dot{\theta}\sin\varphi + a\dot{\psi}\sin\theta\cos\varphi &= v. \end{aligned} \qquad (3.4.4)$$

In a more general problem of this type, we could be faced with the task of solving

$$\frac{\mathrm{d}}{\mathrm{d}t}\left(\frac{\partial T}{\partial v_a}\right) - \frac{\partial T}{\partial q_a} = E_a + K_a \qquad (3.4.5)$$

($a = 1, 2, \ldots, n$) where the E_a are given external forces; and the K_a are unknown constraint forces, responsible for maintaining a number of constraints of the form

$$A_a(q, t)v_a + B(q, t) = 0, \tag{3.4.6}$$

in which the A_a and B are known functions of q_a and t.

It can happen that such a constraint is really a holonomic constraint in disguise. If $f(q, t) = 0$ throughout the motion, then

$$\frac{\mathrm{d}f}{\mathrm{d}t} = \frac{\partial f}{\partial q_a} v_a + \frac{\partial f}{\partial t} = 0 \tag{3.4.7}$$

which is of the same form as eqn (3.4.6). Conversely, if there exists a function $g(q, t)$ such that

$$A_a = g\frac{\partial f}{\partial q_a}, \qquad B = g\frac{\partial f}{\partial t} \tag{3.4.8}$$

for some function $f(q, t)$, then eqn (3.4.6) is equivalent to the condition that f should be constant during the motion; and we can treat $f = \text{constant}$ as a holonomic constraint (for various different values of the constant).

Definition (3.4.1). If eqn (3.4.8) holds, then the constraint equation (3.4.6) is *integrable*; and the function $f(q, t)$ is an *integral* of the constraint.

It can be shown that a necessary and sufficient condition for integrability is that

$$\begin{gathered} F_{ab}A_c + F_{bc}A_a + F_{ca}A_b = 0 \\ BF_{ab} + A_b\frac{\partial A_a}{\partial t} - A_a\frac{\partial A_b}{\partial t} + A_a\frac{\partial B}{\partial q_b} - A_b\frac{\partial B}{\partial q_a} = 0 \end{gathered} \tag{3.4.9}$$

where

$$F_{ab} = \frac{\partial A_b}{\partial q_a} - \frac{\partial A_a}{\partial q_b}. \tag{3.4.10}$$

By using this criterion, one can show that the constraints (3.4.4) are *not* integrable: they are genuinely nonholonomic.

Suppose that we have k constraint equations of the form of eqn (3.4.6):

$$A_{ra}(q, t)v_a + B_r(q, t) = 0 \tag{3.4.11}$$

where $r = 1, 2, \ldots, k$. If the constraints are integrable, with integrals

$f_r(q, t)$, and if the constraint forces are workless, then

$$K_a X_a = 0 \quad \text{whenever} \quad X_a \frac{\partial f_r}{\partial q_a} = 0 \tag{3.4.12}$$

for all r. But this is equivalent to

$$K_a X_a = 0 \quad \text{whenever} \quad X_a A_{ra} = 0 \quad (\text{for all } r), \tag{3.4.13}$$

a condition that makes sense even when the constraints are not integrable.

Definition (3.4.2). The constraint forces K_a are *workless* if $K_a X_a = 0$ whenever $X_a A_{ra} = 0$ for all r.

This may seem a strange condition; its usefulness lies in the fact that it *is* satisfied by the constraint forces in a variety of systems; and, in particular, in most rolling problems. For example, for our sphere, the condition has the following interpretation: freeze the motion of the plane and the sphere, recording, as before, all the forces acting on the system. Choose real numbers X_a $(a = 1, 2, \ldots, 6)$ and consider the small change in the configuration given by $\delta x = \varepsilon X_1$, $\delta y = \varepsilon X_2$, $\delta\theta = \varepsilon X_3$, $\delta\varphi = \varepsilon X_4$, and $\delta\psi = \varepsilon X_5$, where ε is some small parameter and t is held fixed. With the two constraint equations (3.4.4), the condition $X_a A_{ra} = 0$ $(r = 1, 2)$ reads

$$\begin{aligned} X_1 - aX_3 \cos\varphi - aX_5 \sin\theta \sin\varphi &= 0 \\ X_2 - aX_3 \sin\varphi + aX_5 \sin\theta \cos\varphi &= 0; \end{aligned} \tag{3.4.14}$$

or, equivalently,

$$\begin{aligned} \delta x - a\delta\theta \cos\varphi - a\delta\psi \sin\theta \sin\varphi &= 0 \\ \delta y - a\delta\theta \sin\varphi + a\delta\psi \sin\theta \cos\varphi &= 0. \end{aligned} \tag{3.4.14$'$}$$

But this is simply the condition that the displacement should be produced by rolling the sphere a small distance along the plane with the plane held fixed (by reversing the argument that led to eqn (3.4.4)). During such a displacement, the point of contact does not move to the first order in ε—which is simply another way of saying that the displacement is produced by rolling. It follows that if $X_a A_{ra} = 0$ $(r = 1, 2)$, then the friction forces do no work during the displacement and $K_a X_a = 0$. In this case, therefore, the constraint forces are workless in the sense of the definition. Note, however, that this is *not* the same as saying that the friction forces do no work during the *actual* motion—that would be false.

Now consider the values of the functions K_a, A_{ra}, and B at some fixed values of q_a and t.

Proposition (3.4.1). If $X_a K_a = 0$ whenever $X_a A_{ra} = 0$, then there exist real numbers $\lambda_1, \ldots, \lambda_k$ such that

$$K_a = \lambda_1 A_{1a} + \lambda_2 A_{2a} + \cdots + \lambda_k A_{ka}. \tag{3.4.15}$$

Proof. The proposition is a restatement of a standard result in linear algebra.

We shall think of $X_1, X_2, \ldots, X_n$ as the entries in a column vector X. Let A be the $k \times n$ matrix

$$A = \begin{pmatrix} A_{11} & A_{12} & \cdots & A_{1n} \\ A_{21} & A_{22} & \cdots & A_{2n} \\ \vdots & \vdots & & \vdots \\ A_{k1} & A_{k2} & \cdots & A_{kn} \end{pmatrix} \tag{3.4.16}$$

and let ρ be the rank of A: ρ is the dimension of the vector space $V \subset \mathbb{R}^n$ spanned by the rows of A.

Let N be the space of column vectors X such that $AX = 0$. Then the rank-nullity theorem tells us that $\dim(N) = n - \rho$.

Let N^0 be the annihilator of N; that is, N^0 is the space of row vectors

$$C = (c_1, c_2, \ldots, c_n) \tag{3.4.17}$$

such that $CX = 0$ whenever $X \in N$. Then

$$\dim(N^0) = n - \dim(N) = \rho. \tag{3.4.18}$$

Now V is certainly contained in N^0, by the definition of N. But V and N^0 have the same dimension; therefore $V = N^0$.

The condition $K_a X_a = 0$ whenever $A_{ra} X_a = 0$ is simply the condition that the row vector $K = (K_1, \ldots, K_n)$ should lie in N^0. Hence, when it holds, K lies in V and so there exist $\lambda_1, \ldots, \lambda_k \in \mathbb{R}$ such that eqn (3.4.15) holds. □

Suppose now that the constraint forces in eqn (3.4.5) are workless. Then, by applying the proposition at each time during the motion, we can deduce that

$$\frac{\mathrm{d}}{\mathrm{d}t}\left(\frac{\partial T}{\partial v_a}\right) - \frac{\partial T}{\partial q_a} = E_a + \lambda_1 A_{1a} + \cdots + \lambda_k A_{ka} \tag{3.4.19}$$

where $\lambda_1, \ldots, \lambda_k$ are now functions of time. Together with the original constraint equations (3.4.11), this gives a system of $n + k$ equations in the $n + k$ unknown functions of time $q_1(t), \ldots, q_n(t), \lambda_1(t), \ldots, \lambda_k(t)$.

When the external forces are conservative, so that $E_a = -\partial U/\partial q_a$, eqn

(3.4.19) becomes

$$\frac{\mathrm{d}}{\mathrm{d}t}\left(\frac{\partial L}{\partial v_a}\right) - \frac{\partial L}{\partial q_a} = \lambda_1 A_{1a} + \cdots + \lambda_k A_{ka} \tag{3.4.20}$$

where $L = T - U$.

Definition (3.4.3). The functions $\lambda_1(t), \ldots, \lambda_k(t)$ are called *Lagrange multipliers.*

In our example of the sphere rolling on the moving plane, we have seven equations: the two constraints, together with

$$(x)\quad \frac{\mathrm{d}}{\mathrm{d}t}(m\dot{x}) = \lambda$$

$$(y)\quad \frac{\mathrm{d}}{\mathrm{d}t}(m\dot{y}) = \mu$$

$$(\theta)\quad \frac{\mathrm{d}}{\mathrm{d}t}(\tfrac{2}{5}ma^2\dot{\theta}) + \tfrac{2}{5}ma^2\dot{\varphi}\dot{\psi}\sin\theta = -a\lambda\cos\varphi - a\mu\sin\varphi \tag{3.4.21}$$

$$(\varphi)\quad \frac{\mathrm{d}}{\mathrm{d}t}(\tfrac{2}{5}ma^2(\dot{\varphi} + \dot{\psi}\cos\theta)) = 0$$

$$(\psi)\quad \frac{\mathrm{d}}{\mathrm{d}t}(\tfrac{2}{5}ma^2(\dot{\psi} + \dot{\varphi}\cos\theta)) = -a\lambda\sin\theta\sin\varphi + a\mu\sin\theta\cos\varphi$$

where $\lambda = \lambda_1$ and $\mu = \lambda_2$ are the two Lagrange multipliers corresponding to the two constraints. We can see from the first two equations that they are in fact the two components of the friction force at the point of contact.

Exercises

(3.4.1) Rework example (3.3.3) by treating the rolling condition as if it were a nonholonomic constraint.

(3.4.2) A system is subject to workless nonholonomic constraints with $B_r = 0$. Show that if $T = \frac{1}{2}T_{ab}v_a v_b$ where $T_{ab} = T_{ab}(q)$, then

$$\frac{\mathrm{d}T}{\mathrm{d}t} = v_a E_a.$$

(3.4.3)* Solve exercise (3.2.6) by using coordinates and Lagrange multipliers.

4 Hamiltonian mechanics

4.1 Hamilton's equations

In a conservative holonomic system with n degrees of freedom, Lagrange's equations

$$\frac{\mathrm{d}}{\mathrm{d}t}\left(\frac{\partial L}{\partial v_a}\right)-\frac{\partial L}{\partial q_a}=0, \qquad \frac{\mathrm{d}}{\mathrm{d}t}(q_a)=v_a \tag{4.1.1}$$

determine the orbits of the system in the time-phase space PT. We have seen that it is possible to simplify the dynamical analysis by making coordinate transformations of the form

$$\begin{aligned}\tilde{q}_a &= \tilde{q}_a(q, t)\\ \tilde{v}_a &= \tilde{v}_a(q, v, t)=\frac{\partial \tilde{q}_a}{\partial q_b}v_b+\frac{\partial \tilde{q}_a}{\partial t}\\ \tilde{t} &= t,\end{aligned} \tag{4.1.2}$$

a technique that proved particularly useful for handling constraints.

Although eqn (4.1.2) allows arbitrary transformations of the configuration coordinates, the transformation of the velocity coordinates is fixed by the choice made for $\tilde{q}_a=\tilde{q}_a(q, t)$. One might hope to achieve further simplification by allowing transformations that mix up the velocity and configuration coordinates. However, because of the asymmetric way in which the q_a's and v_a's appear in eqn (4.1.1), any general substitution

$$\begin{aligned}\tilde{q}_a &= \tilde{q}_a(q, v, t)\\ \tilde{v}_a &= \tilde{v}_a(q, v, t)\\ \tilde{t} &= t\end{aligned}$$

completely destroys Lagrange's simple form of the equations of motion.

The first step towards introducing more general transformations, therefore, is to replace q_a and v_a by new coordinates that appear more on the same footing in the dynamical equations. The way in which the time derivative of $\partial L/\partial v_a$ enters the first set of equations suggests the following

$$\begin{aligned}\tilde{q}_a &= q_a\\ \tilde{p}_a &= \frac{\partial L}{\partial v_a}\\ \tilde{t} &= t.\end{aligned} \tag{4.1.3}$$

The velocity coordinates v_a are replaced by the generalized momenta $\tilde{p}_a = \tilde{p}_a(q, v, t)$. Together, $\tilde{q}_a$, $\tilde{p}_a$, and $\tilde{t}$ form a set of $2n+1$ coordinates on PT.

As usual, the tildes are used to avoid the second fundamental confusion of calculus: they enable us to distinguish between

$$\frac{\partial}{\partial q_a} \quad \text{and} \quad \frac{\partial}{\partial \tilde{q}_a}. \tag{4.1.4}$$

The first is the partial derivative holding fixed t, the remaining q coordinates, and all the v_a; the second is the partial derivative holding fixed $\tilde{t}$, the remaining $\tilde{q}$ coordinates, and all the $\tilde{p}_a$.

Example (4.1.1). Suppose that $L = \frac{1}{2}v_1^2 - q_1 v_1$ $(n=1)$. Then the new coordinates are $\tilde{q}_1 = q_1$, $\tilde{p}_1 = v_1 - q_1$, and $\tilde{t} = t$. If $f = v_1$, then $\partial f/\partial q_1 = 0$. In the new coordinates, however, $f = \tilde{p}_1 + \tilde{q}_1$ and $\partial f/\partial \tilde{q}_1 = 1$.

A central role in what follows will be played by the *Hamiltonian*, which is the function $h: PT \to \mathbb{R}$ defined by

$$h = \tilde{p}_a v_a - L. \tag{4.1.5}$$

The Hamiltonian can be expressed either as a function of q_a, v_a and t (by substituting for $\tilde{p}_a$ from eqns (4.1.3)); or as a function of $\tilde{q}_a$, $\tilde{p}_a$, and $\tilde{t}$, by inverting eqns (4.1.3) to express the q_a's, v_a's, and t as functions of $\tilde{q}_a$, $\tilde{p}_a$, and $\tilde{t}$.

If we think of h as a function of $\tilde{q}_a$, $\tilde{p}_a$ and $\tilde{t}$, then

$$\begin{aligned}\frac{\partial h}{\partial \tilde{p}_a} &= v_a + \tilde{p}_b \frac{\partial v_b}{\partial \tilde{p}_a} - \frac{\partial L}{\partial q_b}\frac{\partial q_b}{\partial \tilde{p}_a} - \frac{\partial L}{\partial v_b}\frac{\partial v_b}{\partial \tilde{p}_a} - \frac{\partial L}{\partial t}\frac{\partial t}{\partial \tilde{p}_a} \\ &= v_a,\end{aligned} \tag{4.1.6}$$

since

$$\frac{\partial q_b}{\partial \tilde{p}_a} = 0, \qquad \frac{\partial t}{\partial \tilde{p}_a} = 0, \quad \text{and} \quad \frac{\partial L}{\partial v_b} = \tilde{p}_b. \tag{4.1.7}$$

Similarly,

$$\begin{aligned}\frac{\partial h}{\partial \tilde{q}_a} &= \tilde{p}_b \frac{\partial v_b}{\partial \tilde{q}_a} - \frac{\partial L}{\partial q_b}\frac{\partial q_b}{\partial \tilde{q}_a} - \frac{\partial L}{\partial v_b}\frac{\partial v_b}{\partial \tilde{q}_a} - \frac{\partial L}{\partial t}\frac{\partial t}{\partial \tilde{q}_a} \\ &= -\frac{\mathrm{d}\tilde{p}_a}{\mathrm{d}t},\end{aligned} \tag{4.1.8}$$

since

$$\frac{\partial q_b}{\partial \tilde{q}_a} = \delta_{ab}, \qquad \frac{\partial L}{\partial q_b} = \frac{\mathrm{d}}{\mathrm{d}t}\left(\frac{\partial L}{\partial v_b}\right) = \frac{\mathrm{d}\tilde{p}_b}{\mathrm{d}t}, \qquad \frac{\partial t}{\partial \tilde{q}_a} = 0. \tag{4.1.9}$$

Thus in the new coordinates, the equations of motion are

$$\frac{\mathrm{d}\tilde{q}_a}{\mathrm{d}t} = \frac{\partial h}{\partial \tilde{p}_a}, \qquad \frac{\mathrm{d}\tilde{p}_a}{\mathrm{d}t} = -\frac{\partial h}{\partial \tilde{q}_a}, \tag{4.1.10}$$

in which there is an obvious symmetry between the coordinates $\tilde{p}_a$ and $\tilde{q}_a$.

The equations of motion in this form are called *Hamilton's equations* and the coordinate transformation (4.1.3) is called the *Legendre transformation.*

Whittaker[7] points out in his *Analytical dynamics* that the 'Hamiltonian', which appeared in a paper by Hamilton published in 1834, also made an earlier appearance in a paper by Poisson in 1809. Moreover, Poisson obtained half of 'Hamilton's equations'; and the full set appeared in a paper by Lagrange in 1810, in which he considered the effect of a perturbing force on solutions of dynamical problems (the role of h was played by the potential of the disturbing force). The equations also arose in the context of work on the characteristics of first-order partial differential equations in papers by Pfaff (1814–15) and Cauchy (1819). Hamilton's 1834 paper dealt only with the case in which h is a function of $\tilde{p}_a$ and $\tilde{q}_a$, but not $\tilde{t}$. The extension to the time-dependent case was carried out in papers by Ostrogradsky (published 1848–50) and Donkin (1854). The nomenclature in this, as in other parts of classical mechanics, pays little heed to precedence.

In generalized coordinates adapted to holonomic constraints, the Lagrangian of a system of particles with n (residual) degrees of freedom is

$$L = \tfrac{1}{2}K_{ab}v_a v_b + A_a v_a + C, \tag{4.1.11}$$

where K_{ab}, A_a, and C are all functions of the q_a's and t (but not of the v_a's). The generalized momenta are

$$\tilde{p}_a = K_{ab}v_b + A_a. \tag{4.1.12}$$

Hence

$$h = \tfrac{1}{2}K_{ab}v_a v_b - C. \tag{4.1.13}$$

Thus the Hamiltonian is obtained from the Lagrangian by: reversing the signs of the terms that do not involve the velocities; deleting the terms that are linear in the velocities; and leaving unchanged the terms that are quadratic in the velocities.

Combined with the following proposition, this rule of thumb gives a quick way of finding a constant of the motion, which is sometimes helpful in cases in which Hamilton's equations themselves do not open a useful avenue of progress.

Proposition (4.1.1). If $\partial L/\partial t = 0$, then h is a constant of the motion.

Proof. By differentiating h along the orbits in PT, we obtain

$$\begin{aligned}\frac{\mathrm{d}h}{\mathrm{d}t} &= \frac{\partial h}{\partial \tilde{q}_a}\frac{\mathrm{d}\tilde{q}_a}{\mathrm{d}t} + \frac{\partial h}{\partial \tilde{p}_a}\frac{\mathrm{d}\tilde{p}_a}{\mathrm{d}t} + \frac{\partial h}{\partial \tilde{t}} \\ &= \frac{\partial h}{\partial \tilde{t}}\end{aligned} \tag{4.1.14}$$

from Hamilton's equations. But, from eqn (4.1.5),

$$\begin{aligned}\frac{\partial h}{\partial \tilde{t}} &= \frac{\partial v_a}{\partial \tilde{t}}\tilde{p}_a - \frac{\partial L}{\partial \tilde{t}} \\ &= \frac{\partial v_a}{\partial \tilde{t}}\tilde{p}_a - \frac{\partial L}{\partial q_a}\frac{\partial q_a}{\partial \tilde{t}} - \frac{\partial L}{\partial v_a}\frac{\partial v_a}{\partial \tilde{t}} - \frac{\partial L}{\partial t}\frac{\partial t}{\partial \tilde{t}} \\ &= -\frac{\partial L}{\partial q_a}\frac{\partial q_a}{\partial \tilde{t}} - \frac{\partial L}{\partial t}\frac{\partial t}{\partial \tilde{t}}\end{aligned} \tag{4.1.15}$$

since $\tilde{p}_a = \partial L/\partial v_a$. When eqns (4.1.13) are inverted to express q_a, v_a, and t as functions of $\tilde{q}_a$, $\tilde{p}_a$, and $\tilde{t}$, we have

$$\frac{\partial q_a}{\partial \tilde{t}} = 0 \quad \text{and} \quad \frac{\partial t}{\partial \tilde{t}} = 1. \tag{4.1.16}$$

Hence

$$\frac{\partial h}{\partial \tilde{t}} = -\frac{\partial L}{\partial t} \tag{4.1.17}$$

and so $\mathrm{d}h/\mathrm{d}t = 0$ whenever $\partial L/\partial t = 0$. □

The proof illustrates the importance of keeping track of the distinction between the partial derivatives $\partial/\partial t$ and $\partial/\partial \tilde{t}$.

Example (4.1.2). Take $L = \frac{1}{2}v_1^2 + v_1 t$ ($n = 1$). Then $\tilde{p}_1 = v_1 + t$, $\tilde{t} = t$, and $h = \frac{1}{2}v_1^2 = \frac{1}{2}(\tilde{p}_1 - \tilde{t})^2$. We have

$$\frac{\partial h}{\partial \tilde{t}} = -(\tilde{p}_1 - \tilde{t}) = -v_1 = -\frac{\partial L}{\partial t}. \tag{4.1.18}$$

□

We saw in section 2.5 that if $\partial L/\partial t = 0$ and if $L = T - U$ where T is a homogeneous quadratic in the velocities and $U = U(q)$, then the total energy $E = T + U$ is a constant of the motion. But when L has this form, h is also equal to $T + U$, by our rule of thumb. Hence the energy conservation result in section 2.5 is a special case of proposition (4.1.1). It is important to note, however, that, in general, $h \neq E$; and that it is possible for h to be a constant of the motion in a system in which the total energy itself is not conserved (see exercise (4.1.3)).

It is safe to drop the tildes on the coordinates in Hamilton's equations provided that we remember to express h as a function of q_a, $p_a = \partial L/\partial v_a$, and t *before* taking the partial derivative of h with respect to q_a.

Example (4.1.3). Consider the motion of a particle P of mass m moving in the plane under the influence of a force of magnitude $\lambda m/r^2$ directed towards a fixed point O, where r is the distance from O to P. The motion is governed by the Lagrangian

$$L = \tfrac{1}{2}m(\dot{r}^2 + r^2\dot{\theta}^2) + \frac{\lambda m}{r} \tag{4.1.19}$$

where r and θ are polar coordinates with origin O. On putting $q_1 = r$ and $q_2 = \theta$, we have (without the tildes)

$$p_1 = m\dot{r}, \qquad p_2 = mr^2\dot{\theta} \tag{4.1.20}$$

and

$$\begin{aligned} h &= p_1\dot{r} + p_2\dot{\theta} - L \\ &= \tfrac{1}{2}m(\dot{r}^2 + r^2\dot{\theta}^2) - \frac{\lambda m}{r} \\ &= \frac{1}{2m}\left(p_1^2 + \frac{p_2^2}{q_1^2}\right) - \frac{\lambda m}{q_1}. \end{aligned} \tag{4.1.21}$$

In this case Hamilton's equations are

$$\begin{aligned} \dot{q}_1 &= \frac{p_1}{m} \qquad & \dot{q}_2 &= \frac{p_2}{mq_1^2} \\ \dot{p}_1 &= \frac{p_2^2}{mq_1^3} - \frac{\lambda m}{q_1^2} \qquad & \dot{p}_2 &= 0. \end{aligned} \tag{4.1.22}$$

Since $\partial L/\partial t = 0$, the Hamiltonian is a constant of the motion.

Exercises

(4.1.1) Obtain the constant of the motion in exercise (1.3.1) by the Hamiltonian method.

(4.1.2) Obtain the Hamilton's equations for a particle moving in space under an inverse-square-law central force, taking the q_a to be spherical polar coordinates.

(4.1.3)† The ends A and B of a thin uniform rod of mass m and length $2a$ can slide freely: A along a smooth horizontal wire OX and B along a smooth vertical wire OZ, with OZ pointing vertically upwards. The wire frame OXZ is made to rotate with constant angular velocity Ω about OZ.

Show that if B is above O and the angle OBA is θ, then

$$h = \tfrac{2}{3}ma^2(\dot{\theta}^2 - \Omega^2 \sin^2\theta) + mga \cos\theta$$

and that h is a constant of the motion. Is h equal to the total energy?

What is the Hamiltonian if, instead of being rotated about OZ, the frame OXZ is made to rotate with constant angular velocity ω about the horizontal axis OX? Is it conserved?

4.2* Poisson brackets and canonical transformations

Suppose that $f: PT \to \mathbb{R}$. Then

$$\begin{aligned} \frac{\mathrm{d}f}{\mathrm{d}t} &= \frac{\partial f}{\partial q_a}\dot{q}_a + \frac{\partial f}{\partial p_a}\dot{p}_a + \frac{\partial f}{\partial t} \\ &= \frac{\partial f}{\partial q_a}\frac{\partial h}{\partial p_a} - \frac{\partial f}{\partial p_a}\frac{\partial h}{\partial q_a} + \frac{\partial f}{\partial t} \end{aligned} \tag{4.2.1}$$

(we have dropped the tildes on the coordinates q_a, p_a, and t). The combination of q_a and p_a derivatives on the right-hand side is of central importance in the search for coordinate transformations that preserve Hamilton's equations.

Definition (4.2.1). Let f and g be functions on PT. The *Poisson bracket* of f and g is the function

$$[f, g] = \frac{\partial f}{\partial p_a}\frac{\partial g}{\partial q_a} - \frac{\partial f}{\partial q_a}\frac{\partial g}{\partial p_a}. \tag{4.2.2}$$

The Poisson bracket is skew-symmetric; that is $[f, g] = -[g, f]$. And it satisfies the *Jacobi identity*: for any three functions f, g, and k,

$$[f, [g, k]] + [g, [k, f]] + [k, [f, g]] = 0. \tag{4.2.3}$$

The proof is left as an exercise.

When written in terms of the Poisson bracket, Hamilton's equations become

$$\dot{q}_a = [h, q_a] \quad \text{and} \quad \dot{p}_a = [h, p_a] \tag{4.2.4}$$

(by taking f in eqn (4.2.1) to be each of the coordinate functions in turn). These suggest that a good starting point would be to look for coordinate transformations that preserve Poisson brackets.

* This section contains harder material that can be omitted.

Suppose that new coordinates q''_a, p''_a, and t'' are defined on PT by expressions of the form

$$q''_a = q''_a(q, p, t), \qquad p''_a = p''_a(q, p, t), \quad \text{and} \quad t'' = t. \tag{4.2.5}$$

Let $f, g : PT \to \mathbb{R}$ and put

$$[f, g]'' = \frac{\partial f}{\partial p''_a}\frac{\partial g}{\partial q''_a} - \frac{\partial f}{\partial q''_a}\frac{\partial g}{\partial p''_a}. \tag{4.2.6}$$

We want to understand the conditions under which $[f, g] = [f, g]''$ for every f and g.

The key to this problem is the introduction of a third system of coordinates q'_a, p'_a, t', which is intermediate between the unprimed system q_a, p_a, t and the double-primed system q''_a, p''_a, t''. It is defined by

$$q'_a = q_a, \qquad p'_a = p''_a, \qquad t' = t = t''. \tag{4.2.7}$$

The transformation from the unprimed to the double-primed system is thus broken into two stages: first, there is the transformation from q_a, p_a, t to the primed system q'_a, p'_a, t'. This is determined by expressions of the form

$$q_a = q'_a, \qquad p_a = p_a(q', p', t'), \qquad t = t'. \tag{4.2.8}$$

Second, there is the transformation from q'_a, p'_a, t' to q''_a, p''_a, t'', which is given by

$$q''_a = q''_a(q', p', t'), \qquad p''_a = p'_a, \qquad t'' = t'. \tag{4.2.9}$$

We shall use the familiar device to keep track of the distinction between the various partial derivatives: when a partial derivative is taken with respect to a variable of one type (unprimed, primed, or double-primed), the values of the other variables of the *same* type are held fixed. For example, $\partial/\partial q_1$ is the partial derivative with respect to q_1 with $q_2, \ldots, q_n, p_1, \ldots, p_n$, and t held fixed; while $\partial/\partial q'_1$ is the partial derivative with $q'_2, \ldots, q'_n, p'_1, \ldots, p'_n$, and t' held fixed.

The expression

$$\frac{\partial}{\partial q'_1}\left(\frac{\partial f}{\partial q_1}\right) \tag{4.2.10}$$

is legitimate and unambiguous: first express f as a function of q_a, p_a, and t, and take the partial derivative with respect to q_1. Then express the result as a function of q'_a, p'_a, and t' and take the partial derivative with respect to q'_1, with p'_a, t', and the remaining q'_a held fixed.

In general, however,

$$\frac{\partial}{\partial q'_1}\left(\frac{\partial f}{\partial q_1}\right) \neq \frac{\partial}{\partial q_1}\left(\frac{\partial f}{\partial q'_1}\right). \tag{4.2.11}$$

So the notation $\partial^2 f/\partial q_1' \, \partial q_1$ with mixed coordinate systems is very dangerous and should be avoided in this context. But of course partial derivatives with respect to coordinates of the *same* system can be interchanged, so that, for example,

$$\frac{\partial^2 f}{\partial q_1' \, \partial p_1'} = \frac{\partial^2 f}{\partial p_1' \, \partial q_1'}.$$

There is a technical problem, over which we should take some care. It is not obvious that the primed variables will be a good system of coordinates on PT since it may not be possible to express p_a and q_a as smooth functions of q_a', p_a', t'. For example, the double-primed system might be defined by $q_a'' = p_a$, $p_a'' = -q_a$, in which case $q_a' = -p_a'$ everywhere in PT, and the values of the primed variables cannot be used as unambiguous labels for the points of PT. This is a special situation, however; in general, eqns (4.2.8) and (4.2.9) *will* hold and the primed variables can be used as coordinates; in which case we say that the coordinate systems q_a, p_a, t and q_a'', p_a'', t'' are *transversal.*

The implicit function theorem gives as a necessary and sufficient condition for transversality that the two $n \times n$ matrices L and M with entries

$$L_{ab} = \frac{\partial p_b''}{\partial p_a} \quad \text{and} \quad M_{ab} = \frac{\partial q_b}{\partial q_a''} \tag{4.2.12}$$

should be nonsingular everywhere.

Proposition (4.2.1). Suppose that the coordinate system q_a, p_a, t is transversal to the system q_a'', p_a'', t''. Then $[f, g] = [f, g]''$ for all f and g if and only if there exists a function $F: PT \to \mathbb{R}$ such that when F is expressed as a function of q_a', p_a', and t',

$$p_a = \frac{\partial F}{\partial q_a'} \quad \text{and} \quad q_a'' = \frac{\partial F}{\partial p_a'}. \tag{4.2.13}$$

The 'if' part is easy, and it is the only part that will be used. The 'only if' part involves an 'integrability condition' that may not be familiar.

Proof. Let f and g be two functions on PT. When applied to the first transformation (eqn (4.2.8)), the chain rule gives

$$\frac{\partial g}{\partial q_a'} = \frac{\partial g}{\partial q_a} + \frac{\partial g}{\partial p_b}\frac{\partial p_b}{\partial q_a'} \quad \text{and} \quad \frac{\partial f}{\partial p_a'} = \frac{\partial f}{\partial p_b}\frac{\partial p_b}{\partial p_a'} \tag{4.2.14}$$

since

$$\frac{\partial q_b}{\partial q_a'} = \delta_{ab}, \qquad \frac{\partial t}{\partial q_a'} = 0, \qquad \frac{\partial q_b}{\partial p_a'} = 0, \quad \text{and} \quad \frac{\partial t}{\partial p_a'} = 0. \tag{4.2.15}$$

In the same way, we obtain from the second transformation (eqn (4.2.9)),

$$\frac{\partial g}{\partial q'_a} = \frac{\partial g}{\partial q''_b}\frac{\partial q''_b}{\partial q'_a} \quad \text{and} \quad \frac{\partial f}{\partial p'_a} = \frac{\partial f}{\partial p''_a} + \frac{\partial f}{\partial q''_b}\frac{\partial q''_b}{\partial p'_a}. \tag{4.2.16}$$

Hence, by multiplying the first equations in each of (4.2.14) and (4.2.16) by $\partial f/\partial p_a$ and summing over a,

$$\frac{\partial f}{\partial p_a}\left(\frac{\partial g}{\partial q_a} + \frac{\partial g}{\partial p_b}\frac{\partial p_b}{\partial q'_a}\right) = \frac{\partial f}{\partial p_a}\frac{\partial g}{\partial q''_b}\frac{\partial q''_b}{\partial q'_a}. \tag{4.2.17}$$

Similarly, by multiplying the second in each pair by $\partial g/\partial q''_a$ and summing,

$$\left(\frac{\partial f}{\partial p''_a} + \frac{\partial f}{\partial q''_b}\frac{\partial q''_b}{\partial p'_a}\right)\frac{\partial g}{\partial q''_a} = \frac{\partial f}{\partial p_b}\frac{\partial g}{\partial q''_a}\frac{\partial p_b}{\partial p'_a}. \tag{4.2.18}$$

To bring some order into this profusion of derivatives, let A, B, C, and D be the $n \times n$ matrices with entries

$$\begin{aligned} A_{ab} &= \frac{\partial p_a}{\partial q'_b} - \frac{\partial p_b}{\partial q'_a} & B_{ab} &= \frac{\partial q''_b}{\partial q'_a} \\ C_{ab} &= \frac{\partial q''_a}{\partial p'_b} - \frac{\partial q''_b}{\partial p'_a} & D_{ab} &= \frac{\partial p_a}{\partial p'_b}. \end{aligned} \tag{4.2.19}$$

Then, by subtracting from (4.2.17) the same equation with f and g interchanged,

$$[f, g] - A_{ab}\frac{\partial f}{\partial p_a}\frac{\partial g}{\partial p_b} = B_{ab}\left(\frac{\partial f}{\partial p_a}\frac{\partial g}{\partial q''_b} - \frac{\partial g}{\partial p_a}\frac{\partial f}{\partial q''_b}\right). \tag{4.2.20}$$

And similarly, from eqn (4.2.18),

$$[f, g]'' + C_{ab}\frac{\partial f}{\partial q''_a}\frac{\partial g}{\partial q''_b} = D_{ab}\left(\frac{\partial f}{\partial p_a}\frac{\partial g}{\partial q''_b} - \frac{\partial g}{\partial p_a}\frac{\partial f}{\partial q''_b}\right). \tag{4.2.21}$$

For the 'if' part of the proposition, suppose that eqn (4.2.13) holds for some function F on PT. Then

$$A_{ab} = \frac{\partial^2 F}{\partial q'_a\,\partial q'_b} - \frac{\partial^2 F}{\partial q'_b\,\partial q'_a} = 0; \tag{4.2.22}$$

similarly, $C_{ab} = 0$; and

$$B_{ab} = \frac{\partial^2 F}{\partial q'_a\,\partial p'_b} = D_{ab}. \tag{4.2.23}$$

Hence $[f, g] = [f, g]''$ for all f and g.

For the converse, suppose that $[f, g] = [f, g]''$ for all f and g. Then

$$C_{ab}\frac{\partial f}{\partial q''_a}\frac{\partial g}{\partial q''_b} + A_{ab}\frac{\partial f}{\partial p_a}\frac{\partial g}{\partial p_b} + (B_{ab} - D_{ab})\left(\frac{\partial f}{\partial p_a}\frac{\partial g}{\partial q''_b} - \frac{\partial g}{\partial p_a}\frac{\partial f}{\partial q''_b}\right) = 0 \quad (4.2.24)$$

By taking $f = p''_c$, $g = q_d$, we obtain $(B_{ab} - D_{ab})L_{ac}M_{bd} = 0$; that is, $L^T(B - D)M = 0$, which implies that $D = B$ since L and M are nonsingular. Similarly, by taking $f = p_c$, $g = p''_d$, we obtain $A_{cb}L_{bd} = 0$; that is, $AL = 0$, which implies that $A = 0$. Finally, by taking $f = q_c$, $g = q''_d$, we obtain $C = 0$. But $A = C = 0$, $B = D$ are precisely the integrability conditions for the existence of F such that eqn (4.2.13) holds. (For two variables x and y it is easy to show that there exists a function $F(x, y)$ such that

$$\frac{\partial F}{\partial x} = \lambda, \qquad \frac{\partial F}{\partial y} = \mu \quad (4.2.25)$$

for given functions λ and μ of x and y if and only if

$$\frac{\partial \mu}{\partial x} = \frac{\partial \lambda}{\partial y}. \quad (4.2.26)$$

We are simply applying the obvious extension of this to $2n$ variables q_a, p_a.) □

Definition (4.2.2). The transformation from the coordinates q_a, p_a, t to the coordinates q''_a, p''_a, $t'' = t$ is *canonical* if $[f, g] = [f, g]''$ for every f and g. A function $F : PT \to \mathbb{R}$ such that eqn (4.2.13) holds is a *generating function* for the transformation.

The sense in which F generates the coordinate transformation is not entirely straightforward. The first point to note is that it is *not* true that one can construct a canonical transformation by picking an arbitrary function F on PT: in order to use eqn (4.2.13) to obtain the relations between the new double-primed coordinates and the old unprimed coordinates, one must first express F in terms of the intermediate, primed, system of coordinates; and, of course, that is not possible unless one knows the transformation in advance.

Instead, one must set about things in a slightly different order.

(1) Pick a family of functions on CT

$$S = S(q_1, \ldots, q_n, k_1, \ldots, k_n, t) \quad (4.2.27)$$

labelled by n parameters $k_1, k_2, \ldots, k_n$.

(2) Construct functions $p'_1, \ldots, p'_n$ on PT by solving the system of

equations

$$p_a = \left.\frac{\partial S}{\partial q_a}\right|_{k_a = p'_a} \tag{4.2.28}$$

for the p'_a as functions of q_a, p_a, and t (on the right-hand side, differentiate S with respect to q_a, holding the other q coordinates, t, and the parameters k_a constant; and *then* substitute $p'_1, \ldots, p'_n$ for $k_1, \ldots, k_n$ in the resulting expression).

(3) Define the intermediate coordinate system q'_a, p'_a, t' by putting $q'_a = q_a$, and $t' = t$; and define $F : PT \to \mathbb{R}$ by setting

$$F(q'_1, \ldots, q'_n, p'_1, \ldots, p'_n, t') \tag{4.2.29}$$

equal to the value of S at $q_a = q'_a$, $k_a = p'_a$, $t = t'$.

(4) Construct the final coordinate system q''_a, p''_a, t'' by putting

$$p''_a = p'_a, \qquad q''_a = \frac{\partial F}{\partial p'_a}, \qquad t'' = t' = t. \tag{4.2.30}$$

The three coordinate systems—the original unprimed system, the intermediate primed system, and the final double-primed system—are then related by eqns (4.2.7) and (4.2.13), so that the transformation from q_a, p_a, t to q''_a, p''_a, t'' is canonical. There is a sense in which F and S are the same, but it is not the obvious one: S is a family of functions on CT, while F is a function on PT. The 'generating' is really being done by S rather than by F.

There is again a problem over 'transversality': the second step will work only if one can, in fact, invert eqns (4.2.28). By the implicit function theorem, this requires that the $n \times n$ matrix with entries

$$\frac{\partial^2 S}{\partial q_a\, \partial k_b} \tag{4.2.31}$$

should be nonsingular.

Example (4.2.1). With $n = 2$:

(1) Take $S = \frac{1}{2}k_1 q_1^2 + k_2 q_2$. Then

(2) $p_1 = p'_1 q_1$, $p_2 = p'_2$. Hence $p'_1 = p_1/q_1$ and $p'_2 = p_2$.

(3) In terms of the primed coordinates

$$F = \tfrac{1}{2}p'_1 (q'_1)^2 + p'_2 q'_2. \tag{4.2.32}$$

(4) The double-primed coordinates are

$$\begin{aligned} p''_1 &= p'_1 & p''_2 &= p'_2 \\ q''_1 &= \frac{\partial F}{\partial p'_1} = \tfrac{1}{2}(q'_1)^2, & q''_2 &= \frac{\partial F}{\partial p'_2} = q'_2. \end{aligned} \tag{4.2.33}$$

Hence

$$p_1'' = p_1/q_1, \qquad p_2'' = p_2, \qquad q_1'' = \tfrac{1}{2}q_1^2, \qquad q_2'' = q_2. \qquad (4.2.34)$$

Note that $[p_1'', q_1'']'' = 1$, which is consistent with

$$[p_1/q_1, \tfrac{1}{2}q_1^2] = 1. \qquad (4.2.35)$$

Example (4.2.2). The identity transformation has the generating function $F = p_a q_a = p_a' q_a'$ (or, equivalently, $S = k_a q_a$).

Consider a transformation which is close to the identity, with $F = p_a' q_a' + \varepsilon f(q', p', t')$ (ε is a small parameter). Then

$$p_a = p_a' + \varepsilon \frac{\partial f}{\partial q_a'} \quad \text{and} \quad q_a'' = q_a' + \varepsilon \frac{\partial f}{\partial p_a'}, \qquad (4.2.36)$$

so that

$$p_a'' = p_a - \varepsilon \frac{\partial f}{\partial q_a'}, \qquad q_a'' = q_a + \varepsilon \frac{\partial f}{\partial p_a'}. \qquad (4.2.37)$$

Thus the new (double-primed) coordinates differ from the old (unprimed) coordinates by terms of order ε. Hence

$$\frac{\partial f}{\partial q_a'} = \frac{\partial f}{\partial q_a} + \mathrm{O}(\varepsilon) \quad \text{and} \quad \frac{\partial f}{\partial p_a'} = \frac{\partial f}{\partial p_a} + \mathrm{O}(\varepsilon). \qquad (4.2.38)$$

By ignoring terms of order ε^2, therefore,

$$p_a'' = p_a - \varepsilon \frac{\partial f}{\partial q_a}, \qquad q_a'' = q_a + \varepsilon \frac{\partial f}{\partial p_a} \qquad (4.2.39)$$

so that it *is* true that an arbitrary function f on PT generates an *infinitesimal* canonical transformation.

Note that the derivatives of f that appear in eqn (4.2.39) are the same as those in Hamilton's equations. If we take $f = h$, then the infinitesimal canonical transformation is given by moving along the orbits of the system in PT through $\delta t = \varepsilon$. □

Exercises

(4.2.1) With $n = 1$, find the transformation generated by $S = \frac{1}{6}k^2 q^3$ and check that it is canonical.

(4.2.2) With $n = 1$, show that canonical transformations preserve volumes in PT; that is

$$\frac{\partial(q'', p'', t'')}{\partial(q, p, t)} = 1.$$

(4.2.3) With $n=1$, show that if $S=s(kq)$, where s is a function of a single variable, then $p''q''=pq$.

(4.2.4) With $n=2$, find S such that $q_1''=q_2$, $q_2''=q_1$, $p_1''=p_2$, $p_2''=p_1$.

(4.2.5) Show that if $q_a''=q_a''(q)$ and

$$p_a=\frac{\partial q_b''}{\partial q_a}p_b'',$$

then the transformation from q_a, p_a, t to q_a'', p_a'', $t''=t$ is canonical. Show that a generating function is $F=q_a''p_a''$.

4.3* The Hamilton–Jacobi equation

We shall now consider how Hamilton's equations behave under the canonical transformation with generating function $F(q_a', p_a', t')$. As before, q_a, p_a, t is the original coordinate system, q_a'', p_a'', t'', is the transformed system, and q_a', p_a', t' is the intermediate system defined by eqn (4.2.7).

Let G be the function on PT defined by

$$G=\frac{\partial F}{\partial t'}. \tag{4.3.1}$$

Then we have the following.

Proposition (4.3.1). For any function f on PT,

$$\frac{\partial f}{\partial t}=\frac{\partial f}{\partial t''}+[G,f]. \tag{4.3.2}$$

Proof. In the notation of eqns (4.2.19), $A_{ab}=C_{ab}=0$ and $B_{ab}=D_{ab}$. Hence, for any function f on PT,

$$\frac{\partial f}{\partial p_a}\frac{\partial G}{\partial q_a'}=\frac{\partial f}{\partial p_a}B_{ab}\frac{\partial G}{\partial q_b''},\qquad \frac{\partial f}{\partial q_a''}\frac{\partial G}{\partial p_a'}=\frac{\partial f}{\partial q_a''}D_{ba}\frac{\partial G}{\partial p_b}=\frac{\partial f}{\partial q_b''}B_{ab}\frac{\partial G}{\partial p_a} \tag{4.3.3}$$

by using eqns (4.2.16) and (4.2.14).

By applying the chain rule once again to eqn (4.2.8), we obtain

$$\frac{\partial f}{\partial t'}=\frac{\partial f}{\partial p_a}\frac{\partial p_a}{\partial t'}+\frac{\partial f}{\partial t}=\frac{\partial f}{\partial p_a}\frac{\partial G}{\partial q_a'}+\frac{\partial f}{\partial t} \tag{4.3.4}$$

* This section contains harder material that can be omitted.

since $p_a = \partial F/\partial q'_a$. Similarly, from eqn (4.2.9),

$$\frac{\partial f}{\partial t'} = \frac{\partial f}{\partial q''_a}\frac{\partial q''_a}{\partial t'} + \frac{\partial f}{\partial t''} = \frac{\partial f}{\partial q''_a}\frac{\partial G}{\partial p'_a} + \frac{\partial f}{\partial t''} \tag{4.3.5}$$

since $q''_a = \partial F/\partial p'_a$. Hence

$$\begin{aligned}\frac{\partial f}{\partial t} &= \frac{\partial f}{\partial t''} - B_{ab}\left(\frac{\partial f}{\partial p_a}\frac{\partial G}{\partial q''_b} - \frac{\partial f}{\partial q''_b}\frac{\partial G}{\partial p_a}\right) \\ &= \frac{\partial f}{\partial t''} - [f, G],\end{aligned} \tag{4.3.6}$$

where we have used eqn (4.2.20). □

Now consider the time derivative of f along the orbits in PT. We know from eqn (4.2.1) that

$$\frac{\mathrm{d}f}{\mathrm{d}t} = [h, f] + \frac{\partial f}{\partial t}. \tag{4.3.7}$$

But $[h, f] = [h, f]''$ since the transformation is canonical. Hence, by applying the proposition,

$$\frac{\mathrm{d}f}{\mathrm{d}t} = [h, f] + [G, f] + \frac{\partial f}{\partial t''} = [h'', f]'' + \frac{\partial f}{\partial t''}, \tag{4.3.8}$$

where $h'' = h + G$. In particular

$$\dot{q}''_a = \frac{\partial h''}{\partial p''_a}, \qquad \dot{p}''_a = -\frac{\partial h''}{\partial q''_a}. \tag{4.3.9}$$

Therefore *the equations of motion of the system in the new coordinates are again of Hamilton's form, but with Hamiltonian* $h'' = h + G$.

Canonical transformations offer much wider scope for simplifying the equations of motion. In particular, if we can find a transformation such that $h'' = 0$, then the orbits in PT will be given by

$$p''_a = \text{constant} \quad \text{and} \quad q''_a = \text{constant}. \tag{4.3.10}$$

Suppose that the transformation is constructed from a family of functions $S(q, k, t)$ by the sequence of operations set out in section 4.2. Fix, for the moment, the values of the parameters $k_1, \ldots, k_n$ and let Σ be the set of points in PT at which

$$p_a = \frac{\partial S}{\partial q_a} \tag{4.3.11}$$

Then Σ is the $(n+1)$-dimensional surface in PT on which $p'_a = k_a$ (or equivalently, $p''_a = k_a$).

The q_a's and t can be used as coordinates on Σ with (q_a, t) labelling the point with unprimed coordinates

$$q_a, p_a = \frac{\partial S}{\partial q_a}, t \tag{4.3.12}$$

(Fig. 4.3.1). On Σ, h is given as a function of q_a and t by

$$h = h\left(q_1, \ldots, q_n, \frac{\partial S}{\partial q_1}, \ldots, \frac{\partial S}{\partial q_n}, t\right); \tag{4.3.13}$$

in other words, by substituting $\partial S/\partial q_a$ for p_a in the expression for h in the unprimed coordinate system q_a, p_a, t on PT. Similarly, G is given as a function of q_a and t on Σ by

$$G = \frac{\partial S}{\partial t} \tag{4.3.14}$$

since it follows from eqn (4.2.29) that differentiating F with respect to t' with p'_a and q'_a held fixed is equivalent to differentiating S with respect to t with q_a and k_a held fixed.

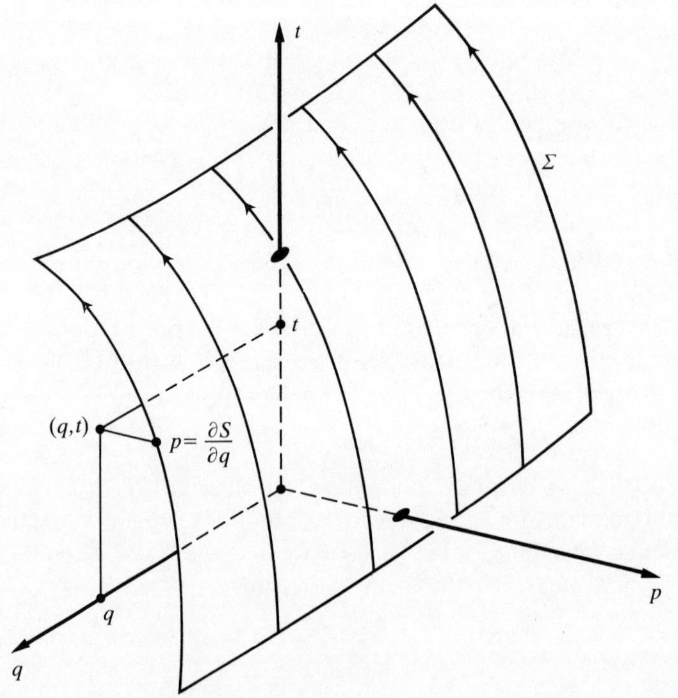

Fig. 4.3.1

Therefore h'' vanishes on Σ if S satisfies the partial differential equation

$$h\left(q_1, \ldots, q_n, \frac{\partial S}{\partial q_1}, \ldots, \frac{\partial S}{\partial q_n}, t\right) + \frac{\partial S}{\partial t} = 0. \tag{4.3.15}$$

This is the *Hamilton–Jacobi equation.*

As the k_a's vary, the corresponding surfaces Σ fill out PT. We can conclude, therefore, that *a family of solutions $S(q, k, t)$ of the Hamilton–Jacobi equation gives rise to a canonical transformation for which h'' vanishes.*

The catch is that it is generally much harder to solve the Hamilton–Jacobi equation, which is a first-order nonlinear partial differential equation, than it is to solve Hamilton's equations. In fact, except in very special systems, the only generally applicable method is the 'method of characteristics', which is based on the reverse construction, the first step being the return to Hamilton's equations.

For this reason, canonical transformations and the Hamilton–Jacobi equation are rarely of practical use for finding *analytic* solutions to mechanical problems. They are, nevertheless, very important for a number of reasons. For example, they provide the connecting link between classical and quantum mechanics (through the JWKB (or WKB) approximation); they provide significant insights into the *qualitative* behaviour of systems that defy analytical treatment; and they play a central part in the theory of partial differential equations.

Even when one can find an n-parameter family of solutions, the best way to proceed is not always to carry out the canonical transformation explicitly. If Σ is the surface in PT corresponding to a particular set of values for the parameters k_a, then $p''_a = p'_a = k_a =$ constant on Σ. But the orbits in PT are given in the double-primed coordinates by eqn (4.3.10). It follows that an orbit that passes through one point of Σ lies in Σ. Moreover, the coordinates q''_a are also constant along the orbits; and, on Σ, q''_a is given as a function of the q_a's and t by

$$q''_a = \frac{\partial S}{\partial k_a} \tag{4.3.16}$$

(by combining eqns (4.2.13) and (4.2.29)). Hence the orbits that lie in Σ are the curves

$$\partial S / \partial k_a = \text{constant} \tag{4.3.17}$$

(in the coordinate system q_a, t on Σ).

By rewriting these in the form $q_a = q_a(t)$, one obtains a family of solutions of the original dynamical problem (the members of the family are labelled by the constants on the right-hand side).

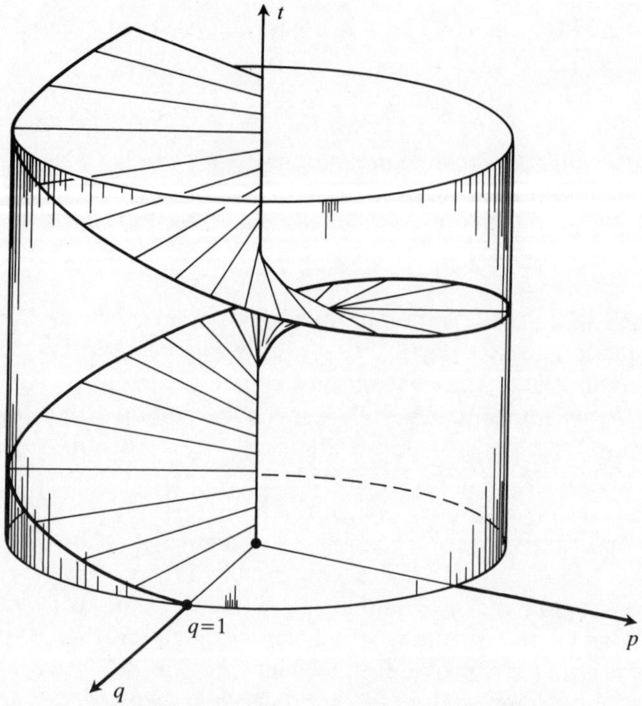

Fig. 4.3.2 The cylinder is the surface $p'' = \frac{1}{2}$; the helicoid is $q'' = \pi/2$. Their intersection (the helix through $(q, p, t) = (1, 0, 0)$) is an orbit of the system in PT.

Example (4.2.1) The harmonic oscillator. Consider the Hamiltonian of the harmonic oscillator

$$h = \tfrac{1}{2}p^2 + \tfrac{1}{2}q^2 \tag{4.3.18}$$

(with $m = 1$).

The Hamilton–Jacobi equation is

$$\frac{\partial S}{\partial t} + \frac{1}{2}\left(\frac{\partial S}{\partial q}\right)^2 + \tfrac{1}{2}q^2 = 0. \tag{4.3.19}$$

It has a one-parameter family of solutions $S = -kt + g(q)$ where

$$g(q) = \int (2k - q^2)^{1/2}\, \mathrm{d}q; \tag{4.3.20}$$

the constant k labels the different solutions in the family.

We can construct a canonical transformation from S by applying the instructions set out in section 4.2. The intermediate coordinates are given by

$$q = q', \qquad t = t'$$
$$p = \left.\frac{\partial S}{\partial q}\right|_{k=p'} = (2p' - q^2)^{1/2} \tag{4.3.21}$$

and the generating function is

$$F = S\,|_{k=p'} = -p't' + \int (2p' - q'^2)^{1/2}\,\mathrm{d}q'. \tag{4.3.22}$$

Therefore double-primed coordinates (Fig. 4.3.2) are

$$p'' = p' = \tfrac{1}{2}(p^2 + q^2)$$
$$q'' = \frac{\partial F}{\partial p'} = -t' + \int \frac{\mathrm{d}q'}{(2p' - q'^2)^{1/2}} = -t + \sin^{-1}(q/\sqrt{(2p')}). \tag{4.3.23}$$

In this case, PT is $\mathbb{R}^3$ (with coordinates q, p, t) and it is easy to solve the equations of motion: the orbits in PT are the helices

$$q = \sqrt{(2E)}\sin(t + \varepsilon), \qquad p = \sqrt{(2E)}\cos(t + \varepsilon) \tag{4.3.24}$$

where ε and E are constants.

Let r and θ be the polar coordinates defined by

$$p = r\cos\theta, \qquad q = r\sin\theta. \tag{4.3.25}$$

Then the orbits can be written more simply as

$$r = \sqrt{(2E)}, \qquad \theta = t + \varepsilon. \tag{4.3.26}$$

In terms of r and θ, the new canonical coordinates are

$$p'' = \tfrac{1}{2}r^2, \qquad q'' = \theta - t, \qquad t'' = t. \tag{4.3.27}$$

As the theory predicts, the orbits are $p'' =$ constant, $q'' =$ constant.

Example (4.3.2) Inverse-square-law central force. In the inverse-square-law problem (example (4.1.1)), the Hamiltonian is

$$h = \frac{1}{2m}(p_1^2 + p_2^2/q_1^2) - \frac{\lambda m}{q_1} \tag{4.3.28}$$

where $q_1 = r$ and $q_2 = \theta$.

The Hamilton–Jacobi equation is

$$\frac{\partial S}{\partial t} + \frac{1}{2m}\left(\frac{\partial S}{\partial q_1}\right)^2 + \frac{1}{2mq_1^2}\left(\frac{\partial S}{\partial q_2}\right)^2 - \frac{\lambda m}{q_1} = 0. \tag{4.3.29}$$

We shall solve this by separating the variables. That is, we shall look for solutions of the form

$$S = f(t) + g_1(q_1) + g_2(q_2). \tag{4.3.30}$$

On substituting into eqn (4.3.29),

$$-\frac{df}{dt} = \frac{1}{2m}\left(\frac{dg_1}{dq_1}\right)^2 + \frac{1}{2mq_1^2}\left(\frac{dg_2}{dq_2}\right)^2 - \frac{\lambda m}{q_1}.$$

The left-hand side is a function of t alone; and the right-hand side of q_1 and q_2 alone. Therefore both sides must be equal to a constant, which we shall denote k_1.

Thus we have $f = -k_1 t + \text{constant}$ and

$$\frac{dg_2}{dq_2} = q_1\left[2mk_1 - \left(\frac{dg_1}{dq_1}\right)^2 + \frac{2\lambda m^2}{q_1}\right]^{1/2}. \tag{4.3.31}$$

Again both sides must be equal to a constant, which we shall denote k_2. Hence

$$g_2 = k_2 q_2 + \text{constant}, \qquad \frac{dg_1}{dq_1} = F(q_1) \tag{4.3.32}$$

where

$$F(q_1) = \left[\frac{2\lambda m^2}{q_1} - \frac{k_2^2}{q_1^2} + 2k_1 m\right]^{1/2}. \tag{4.3.33}$$

Therefore

$$S(q, k, t) = -k_1 t + k_2 q_2 + \int F\, dq_1 \tag{4.3.34}$$

is a two-parameters family of solutions (there is an additional additive constant of integration, which has been dropped since it has no effect on the transformation generated by S).

Rather than look directly at the canonical transformation (which is complicated), we shall use eqn (4.3.17) to deduce that the orbits are given by

$$\frac{\partial S}{\partial k_1} = -t + \int \frac{m\, dq_1}{F} = \text{constant} \tag{4.3.35}$$

and

$$\frac{\partial S}{\partial k_2} = q_2 - \int \frac{k_2\, dq_1}{q_1^2 F} = \text{constant} \tag{4.3.36}$$

for the different values of k_1 and k_2.

These become more familiar when we restore $q_1 = r$ and $q_2 = \theta$ and remember that

$$\tfrac{1}{2}m\left(\dot{r}^2 + \frac{k^2}{r^2}\right) - \frac{\lambda m}{r} = E \tag{4.3.37}$$

where E and $k = r^2\dot{\theta}$ are constant. If we identify k_2 with mk and k_1 with E, then

$$\dot{r} = \frac{1}{m}F(r), \qquad \frac{\mathrm{d}\theta}{\mathrm{d}r} = \frac{km}{r^2F(r)}, \tag{4.3.38}$$

which integrate to give eqns (4.3.35) and (4.3.36).

The Hamilton–Jacobi theory has provided a sophisticated derivation of a familiar and straightforward result, which is typical of its track record as an analytical technique. There are occasional exceptions, however: for example, it is an invaluable aid in the integration of geodesic equations in general relativity; in particular in the Kerr space–time, which represents the gravitational field of a rotating black hole.

Exercises

(4.3.1) Consider a system with one degree of freedom and Hamiltonian $h = h(q, p, t)$. Show that the orbits in $PT = \mathbb{R}^3$ are tangent to the vector field

$$\boldsymbol{X} = \frac{\partial h}{\partial p}\boldsymbol{i} - \frac{\partial h}{\partial q}\boldsymbol{j} + \boldsymbol{k}$$

where $\boldsymbol{i}$, $\boldsymbol{j}$, and $\boldsymbol{k}$ are unit vectors along the q, p, and t axes.

Let Σ be a surface in $\mathbb{R}^3$ given by

$$p = \frac{\partial S}{\partial q}$$

where $S = S(q, t)$. Show that if S is a solution of the Hamilton–Jacobi equation, then $\boldsymbol{X}$ is tangent to Σ. Show conversely that if $\boldsymbol{X}$ is tangent to Σ, then

$$\frac{\partial}{\partial q}\left(\frac{\partial S}{\partial t} + h\left(q, \frac{\partial S}{\partial q}, t\right)\right) = 0.$$

(4.3.2) Solve the Hamilton–Jacobi equation by separating the variables for a particle moving in space under an inverse-square-law central force, taking the q_a to be spherical polar coordinates.

(4.3.3) A particle P of mass m is moving in the plane under the influence of two inverse-square-law forces: $\lambda(PA)^{-2}$ directed towards the point A and $\lambda(PB)^{-2}$ directed towards the point B, where A and B are separated by a distance $2b$. Solve the Hamilton–Jacobi equation in the coordinates θ and φ, where $2b \cosh\varphi = PA + PB$ and $2b\cos\theta = PA - PB$.

5 Impulses

5.1 Generalized impulses

Consider a system of N particles with masses m_α ($\alpha = 1, 2, \ldots, N$). Suppose that the system is subject to holonomic constraints maintained by workless constraint forces. Then the equations of motion are

$$m_\alpha \ddot{\boldsymbol{r}}_\alpha = \boldsymbol{K}_\alpha + \boldsymbol{E}_\alpha \tag{5.1.1}$$

where the $\boldsymbol{K}_\alpha$ are the constraint forces and the $\boldsymbol{E}_\alpha$ are the external forces. As in section 2.5, we can eliminate the $\boldsymbol{K}_\alpha$ by introducing generalized coordinates adapted to the constraints. The result is n second-order equations of motion (where n is now the number of residual degrees of freedom) involving only the external forces:

$$\frac{\mathrm{d}}{\mathrm{d}t}\left(\frac{\partial T}{\partial v_a}\right) - \frac{\partial T}{\partial q_a} = E_a. \tag{5.1.2}$$

(The tildes on the q_a have been dropped, but otherwise the notation is the same as in section 2.5.)

When the constraints are satisfied, the position vector $\boldsymbol{r}_\alpha$ of particle α can be expressed as a function of t and of the free, unconstrained coordinates q_a ($a = 1, 2, \ldots, n$, with summation and range conventions).

If $\delta q_a = \varepsilon X_a$ is a small displacement consistent with the constraints (with t held fixed), then

$$E_a X_a = \sum_\alpha \boldsymbol{E}_\alpha \,.\, \boldsymbol{X}_\alpha, \quad \text{where} \quad \boldsymbol{X}_\alpha = \frac{\partial \boldsymbol{r}_\alpha}{\partial q_a} X_a. \tag{5.1.3}$$

Now suppose that the external forces are impulsive; that is, they are idealized limits of large forces $\boldsymbol{E}_\alpha$ that act for a short time between t and $t + \delta t$. The limit is taken in such a way that the *vector impulses*

$$\boldsymbol{J}_\alpha = \int_t^{t+\delta t} \boldsymbol{E}_\alpha \, \mathrm{d}t \tag{5.1.4}$$

remain finite. Then the *generalized impulses*

$$J_a = \int_t^{t+\delta t} E_a \, \mathrm{d}t \tag{5.1.5}$$

will also remain finite in the limit $\delta t \to 0$.

On integration from t to $t + \delta t$, eqn (5.1.2) becomes

$$\left[\frac{\partial T}{\partial v_a}\right]_t^{t+\delta t} = J_a + \int_t^{t+\delta t} \frac{\partial T}{\partial q_a}\,\mathrm{d}t. \tag{5.1.6}$$

Now the quantities $\partial T/\partial q_a$ are functions on the time–phase space PT: they depend on the configuration and state of motion of the particles, and on t, but not on the particular external forces that are acting on the system. As we take the limit $\delta t \to 0$, the orbit of the system in PT between t and $t + \delta t$ will remain in a bounded region of PT. Therefore the functions $\partial T/\partial q_a$ also remain bounded as $\delta t \to 0$ and the integral on the right-hand side vanishes in the limit. Thus in the limit the integrated equations of motion reduce to

$$\Delta p_a = J_a \tag{5.1.7}$$

where $p_a = \partial T/\partial v_a$ is the generalized momentum conjugate to q_a and Δp_a is the limit of $p_a(t + \delta t) - p_a(t)$. The effect of the impulses is to produce an instantaneous change in the generalized momenta, with the configuration itself remaining unchanged.

To put the generalized impulse equations to work, we need to be able to find the J_a in terms of the vector impulses. The following observation provides the key.

When the constraints are satisfied, the velocity of particle α is

$$\boldsymbol{v}_\alpha = \frac{\partial \boldsymbol{r}_\alpha}{\partial q_a} v_a + \frac{\partial \boldsymbol{r}_\alpha}{\partial t}. \tag{5.1.8}$$

If we keep the configuration and t fixed, but change the v_a's by replacing v_a by $v_a + k_a$, then $\boldsymbol{v}_\alpha$ is changed to $\boldsymbol{v}_\alpha + \boldsymbol{k}_\alpha$, where

$$\boldsymbol{k}_\alpha = \frac{\partial \boldsymbol{r}_\alpha}{\partial q_a} k_a. \tag{5.1.9}$$

But this is the same as the relationship between $X_1, \ldots, X_n$ and the vectors $\boldsymbol{X}_\alpha$ in eqn (5.1.3). Therefore

$$E_a k_a = \sum_\alpha \boldsymbol{k}_\alpha \cdot \boldsymbol{E}_\alpha \tag{5.1.10}$$

and so

$$J_a k_a = \sum_\alpha \boldsymbol{k}_\alpha \cdot \boldsymbol{J}_\alpha. \tag{5.1.11}$$

The generalized impulses can therefore be found by the following procedure. There is no need to bring in the q_a; all that one has to do is:

(1) Write down velocities $v_1, v_2, \ldots, v_n$ which, first, are sufficient to determine the state of motion when the system is in the configuration in which the impulses are applied; and, second, can be assigned values independently of each other without violating the constraints.
(2) Write down the expression for T in terms of the velocities in this configuration.
(3) Find the J_a's by varying v_a and using

$$J_a \delta v_a = \sum \boldsymbol{J} \cdot \delta \boldsymbol{v} \tag{5.1.12}$$

where the sum is over the points P at which the impulses are applied and $\delta \boldsymbol{v}$ is the corresponding variation in the velocity of P. For example, take $\delta v_a = 0$ for $a > 1$. Then the left-hand side is $J_1 \delta v_1$, and eqn (5.1.12) gives J_1 in terms of the $\boldsymbol{J}$'s.
(4) Apply the impulse equations (5.1.7) to find the change in the state of motion produced by the impulses.

It is always possible to find generalized coordinates q_a such that the chosen v_a are the corresponding generalized velocities *in one particular configuration.* But if, for example, the v_a are the three components of the angular velocity $\boldsymbol{\omega}$ of a rigid body, then it is not possible to find coordinates for which the v_a are the generalized velocities in *every* configuration. In impulse problems, however, one considers only one configuration and so it is quite legitimate to take the v_a in step (1) to be the components of $\boldsymbol{\omega}$.

Example (5.1.1). A uniform rod AB of mass m and length a lies at rest on a smooth horizontal plane. It is struck at a point P, distance sa from A $(0 \leqslant s \leqslant 1)$, by an impulse J orthogonal to the rod. The problem is to find the velocities of A and B immediately after the impulse (Fig. 5.1.1).

Let $\boldsymbol{j}$ be a unit vector along AB and let $\boldsymbol{i}$ be a unit vector orthogonal to AB and in the same direction as the impulse.

(1) Let $\boldsymbol{v}_A$ and $\boldsymbol{v}_B$ be the velocities of A and B. The state of motion in the initial configuration is fixed by $v_1 = \boldsymbol{i} \cdot \boldsymbol{v}_A$, $v_2 = \boldsymbol{j} \cdot \boldsymbol{v}_A$, and $v_3 = \boldsymbol{i} \cdot \boldsymbol{v}_B$. Moreover any set of values for v_1, v_2, and v_3 determines a possible state of motion in this configuration.
(2) The kinetic energy is

$$\begin{aligned} T &= \tfrac{1}{6} m (\boldsymbol{v}_A \cdot \boldsymbol{v}_A + \boldsymbol{v}_B \cdot \boldsymbol{v}_B + \boldsymbol{v}_A \cdot \boldsymbol{v}_B) \\ &= \tfrac{1}{6} m (v_1^2 + 3v_2^2 + v_3^2 + v_1 v_3). \end{aligned} \tag{5.1.13}$$

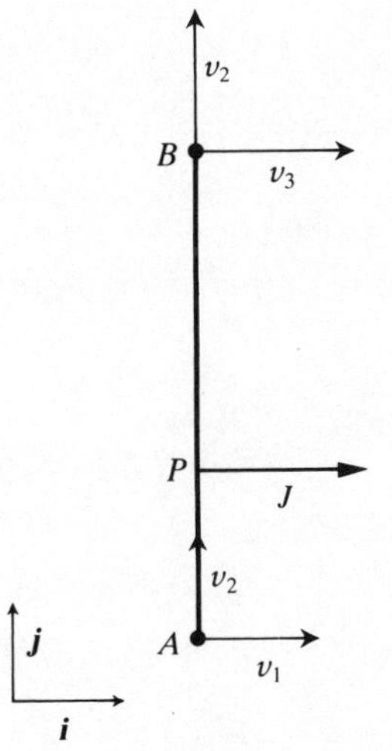

Fig. 5.1.1

(3) The velocity of the point of application (P) is

$$((1-s)v_1+sv_3)\boldsymbol{i}+v_2\boldsymbol{j} \tag{5.1.14}$$

and the impulse at P is $\boldsymbol{J}=J\boldsymbol{i}$. Therefore

$$\begin{aligned} J_1\delta v_1 &= (1-s)\delta v_1 J \\ J_2\delta v_2 &= 0 \\ J_3\delta v_3 &= s\delta v_3 J, \end{aligned} \tag{5.1.15}$$

giving $J_1=(1-s)J$, $J_2=0$, $J_3=sJ$.
(4) The impulse equations are

$$\begin{aligned} p_1 &= \frac{\partial T}{\partial v_1} = \frac{m}{3}v_1+\frac{m}{6}v_3=(1-s)J \\ p_2 &= \frac{\partial T}{\partial v_2} = mv_2 = 0 \\ p_3 &= \frac{\partial T}{\partial v_3} = \frac{m}{3}v_3+\frac{m}{6}v_1=sJ \end{aligned} \tag{5.1.16}$$

giving

$$\begin{aligned} v_1 &= (4-6s)\frac{J}{m} \\ v_2 &= 0 \\ v_3 &= (6s-2)\frac{J}{m} \end{aligned} \tag{5.1.17}$$

immediately after the application of the impulse. □

Suppose that $T=\frac{1}{2}T_{ab}v_a v_b$, that $v_a=u_a$ immediately before the application of the impulses, and that $v_a=w_a$ immediately afterwards. Then $p_a=T_{ab}v_b$ and

$$T_{ab}(w_b-u_b)=J_a. \tag{5.1.18}$$

Let $K_J(u_a)$ be the increment in the kinetic energy generated by the impulses. Then

$$\begin{aligned} K_J(u_a) &= \tfrac{1}{2}T_{ab}(w_a w_b - u_a u_b) \\ &= \tfrac{1}{2}T_{ab}w_a(w_b-u_b)+\tfrac{1}{2}T_{ab}u_a(w_b-u_b) \\ &= \tfrac{1}{2}J_a(w_a+u_a). \end{aligned} \tag{5.1.19}$$

If we replace u_a by u_a+k_a (keeping the J_a's fixed), then w_a is replaced by w_a+k_a. Therefore

$$K_J(u_a+k_a)=J_a k_a+K_J(u_a) \tag{5.1.20}$$

and hence $J_a=\partial K_J/\partial u_a$, which gives another interpretation of J_a.

Example (5.1.2).† Three uniform rods AB, BC, and CD, each of mass m, are at rest on a smooth horizontal table. They are smoothly jointed at B and C and they are arranged to form an open square. An impulse J is applied at A in the direction AC. The problem is to find the kinetic energy of the subsequent motion.

In this configuration, the motion of the system is completely determined by

(1) The two components x and y of the velocity of A;
(2) The component z along BC of the velocity of B; and
(3) The two components u and v of the velocity of D.

Moreover, x, y, z, u, and v can be specified independently. (See Fig. 5.1.2; the notation $x, y, \ldots$ is more convenient for practical calculation—but not for theoretical analysis—than the subscript notation $v_1, v_2, \ldots$.)

The total kinetic energy is

$$T=\tfrac{1}{6}m(x^2+4y^2+5z^2+4v^2+u^2+xz+yv+uz) \tag{5.1.21}$$

(see exercise (3.1.4)).

Write $J_x, J_y, \ldots$ for the generalized impulses. Then, by making variations in x, y, z, u, and v successively

$$\begin{aligned} J_x\delta x &= \delta x\boldsymbol{i}\,.\,\boldsymbol{J}=\frac{1}{\sqrt{2}}J\delta x \\ J_y\delta y &= \delta y\boldsymbol{j}\,.\,\boldsymbol{J}=\frac{1}{\sqrt{2}}J\delta y \\ J_z\delta z &= J_u\delta u = J_v\delta v = 0 \end{aligned} \tag{5.1.22}$$

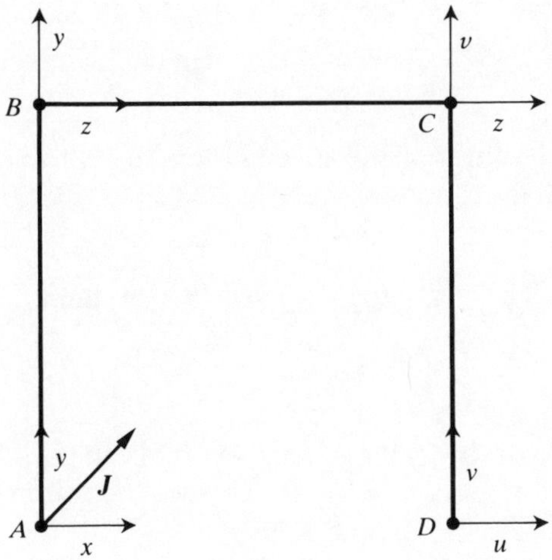

Fig. 5.1.2

where $\boldsymbol{i}$ and $\boldsymbol{j}$ are unit vectors along AD and AB. Hence $J_x = J_y = J/\sqrt{2}$ and $J_z = J_u = J_v = 0$. Therefore the velocities immediately after the impulse are given by

$$(x)\ \frac{m}{6}(2x+z) = \frac{J}{\sqrt{2}} \qquad (y)\ \frac{m}{6}(8y+v) = \frac{J}{\sqrt{2}}$$

$$(z)\ \frac{m}{6}(10z+x+u) = 0 \qquad (u)\ \frac{m}{6}(2u+z) = 0 \tag{5.1.23}$$

$$(v)\ \frac{m}{6}(8v+y) = 0.$$

By solving these,

$$x = \frac{19}{6\sqrt{2}}\frac{J}{m}, \qquad y = \frac{16}{21\sqrt{2}}\frac{J}{m}. \tag{5.1.24}$$

We could also find z, u, and v, and so find T by substitution into (5.1.21). It is simpler, however, to use eqn (5.1.19), which gives the kinetic energy after the impulse as

$$T = \tfrac{1}{2}(xJ_x + yJ_y) = \frac{55}{56}\frac{J^2}{m}. \tag{5.1.25}$$

□

The same technique can also be used to solve initial motion problems. Suppose that a system with $T = \frac{1}{2}T_{ab}(q)v_a v_b$ is released from rest under the influence of external forces with q-components E_a. Then $v_a = 0$ in the initial configuration and

$$T_{ab}\dot{v}_b = E_a. \tag{5.1.26}$$

To find the initial values of the accelerations $\dot{v}_a$, all that we need do is: follow steps (1) and (2) above; then

(3′) Find the E_a's by making variations δv_a in the v_a's and writing

$$E_a \delta v_a = \sum \boldsymbol{E} \cdot \delta \boldsymbol{v} \tag{5.1.27}$$

where the sum is over the points of application of the external forces.

(4′) Solve eqn (5.1.26) for the $\dot{v}_a$'s.

Example (5.1.3) An initial motion problem. Two equal uniform rods AB and BC, each of mass m, are smoothly jointed at B and can move in a vertical plane. The point A is fixed and AB is free to turn about A. Initially the rods are held in a horizontal line and then released. The problem is to find the initial acceleration of C.

(1) Let v_1 and v_2 be the velocities of B and C in the downward vertical direction (Fig. 5.1.3).

(2) The kinetic energy in the initial configuration is

$$T = \frac{m}{6} v_1^2 + \frac{m}{6}(v_1^2 + v_2^2 + v_1 v_2). \tag{5.1.28}$$

(3′) The external forces are the gravitational forces, which act through the centres D and F of AB and BC. The downward velocities of D and F are $\frac{1}{2}v_1$ and $\frac{1}{2}(v_1 + v_2)$ respectively. Hence

$$\begin{aligned} E_1 \delta v_1 &= \tfrac{1}{2} mg \delta v_1 + \tfrac{1}{2} mg \delta v_1 \\ E_2 \delta v_2 &= \tfrac{1}{2} mg \delta v_2 \end{aligned} \tag{5.1.29}$$

which give $E_1 = mg$ and $E_2 = \frac{1}{2}mg$.

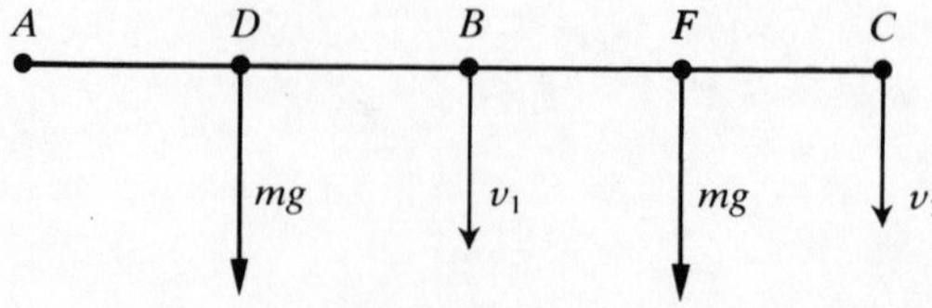

Fig. 5.1.3

(4′) The initial values of $\dot{v}_1$ and $\dot{v}_2$ are given by

$$\frac{m}{3}\dot{v}_1+\frac{m}{6}(2\dot{v}_1+\dot{v}_2)=mg,\quad \frac{m}{6}(2\dot{v}_2+\dot{v}_1)=\tfrac{1}{2}mg \qquad (5.1.30)$$

Hence $\dot{v}_1=9g/7$ and $\dot{v}_2=6g/7$. The initial acceleration of C is therefore $6g/7$ (downwards).

Exercises

(5.1.1)† Two uniform rods AB, BC, each of mass m and length $2a$, are freely jointed at B and lie at rest on a smooth horizontal table. At its midpoint, the rod AB passes through a small smooth ring R which permits the rod AB to move freely along its length and to turn freely about R. Initially the rods are perpendicular and an impulse J is applied at C in the direction CA. Find the motion of the system immediately after the impulse.

(5.1.2)† Three identical uniform rods OX, OY, and OZ, each of mass m, are freely pivoted at O, which is not a fixed point. The rods are orthogonal to each other when an impulse J is applied at X in the direction YX. Show that the kinetic energy generated by the impulse is $23J^2/24m$.

(5.1.3)† Five uniform rods AB, BC, CD, DE, and EA, each of mass m, have lengths $a\sqrt{2}$, $2a$, $2a$, $2a$, $a\sqrt{2}$ respectively. They are freely jointed together to form a pentagon and they lie at rest on a smooth horizontal table, with BC, CD, and DE forming three sides of a square. The perpendicular AP from A to CD is of length $3a$.

An impulse of magnitude J is applied at A in the direction AP. Show that the velocity of A immediately afterwards is $5J/9m$.

(5.1.4)[9] Four equal uniform rods OA, AB, BC, CO are freely jointed to each other at A, B, and C. The rods OA and CO are freely pivoted at the fixed point O. Initially the rods are held so that they form a square in a vertical plane with B vertically below O.

Show that if the rods are released from rest, then the initial acceleration of B will have magnitude $6g/5$.

(5.1.5)† Four uniform rods AB, BC, CD, and DA, each of mass m and length $2a$, are smoothly jointed at A, B, C, and D. The system can rotate freely about A, which is fixed, on a smooth horizontal plane. The rods lie at rest in the form of a square. An impulse J is applied at C in the direction parallel to DB. Show that the initial angular velocity ω of each rod is $3\sqrt{2}J/20ma$.

Show that if ψ is the angle DAB, then immediately after the impulse, $5\ddot{\psi}+6\omega^2=0$.

(5.1.6) A rigid body is free to rotate about a fixed point O. It is set in motion by an impulse $\boldsymbol{J}$ applied at a point with O-position vector $\boldsymbol{r}$. Show that if the initial states of motion are labelled by $v_1 = \omega_1$, $v_2 = \omega_2$, and $v_3 = \omega_3$, then the corresponding generalized impulses are the three components of $\boldsymbol{r} \wedge \boldsymbol{J}$.

(5.1.7)† The top described in examples (3.2.2) and (3.3.1) is precessing steadily about the vertical with angular speed Ω at an acute angle α to the vertical when it is subjected to an impulse J acting through its centre of mass normal to the vertical plane containing the axis of symmetry and in the direction of φ increasing. Show that θ decreases initially (that is, immediately after the impulse) if

$$J^2 a \cot \alpha - CnJ + 2A\Omega J \cos \alpha < 0.$$

5.2 Kelvin's theorem

We shall now specialize to the case in which the holonomic constraints on the system are non-moving; and the kinetic energy is $T = \frac{1}{2}T_{ab}v_a v_b$, where the T_{ab}'s are the entries in a positive definite symmetric matrix T (which can vary with q_a).

Suppose that the system is set in motion from rest by impulses $\boldsymbol{J}_1, \ldots, \boldsymbol{J}_k$ applied at points with position vectors $\boldsymbol{r}_1, \ldots, \boldsymbol{r}_k$; and that while we have not been told the actual values of these impulses, we do know the velocities $\boldsymbol{u}_1, \ldots, \boldsymbol{u}_k$ of the points of application immediately afterwards. There is then a pleasing 'principle of laziness' that enables us to find the state of motion without actually calculating $\boldsymbol{J}_1, \ldots, \boldsymbol{J}_k$; it goes by the name of *Kelvin's theorem*.

Introduce generalized coordinates adapted to the constraints and suppose that the generalized velocities initially (that is, immediately after the application of the impulses) are v_a; as before, $a = 1, 2, \ldots, n$, where n is the number of residual degrees of freedom. Let $\hat{v}_a$ be any other set of values for the generalized velocities for which the points $P_1, \ldots, P_k$ have the same velocities $\boldsymbol{u}_1, \ldots, \boldsymbol{u}_k$ as in the actual motion. Then Kelvin's theorem states that

$$\tfrac{1}{2}T_{ab}v_a v_b \leqslant \tfrac{1}{2}T_{ab}\hat{v}_b\hat{v}_b. \tag{5.2.1}$$

In other words, the kinetic energy in the actual motion takes the lowest possible value consistent with the constraints and the prescribed velocities of the points $P_1, \ldots, P_k$.

To prove the theorem, let the generalized impulses be J_a. Since $\partial \boldsymbol{r}_\alpha / \partial t = 0$ in this case, eqn (5.1.11) gives

$$J_a v_a = \sum \boldsymbol{J}_\alpha \cdot \boldsymbol{v}_\alpha \tag{5.2.2}$$

by taking $k_a = v_a$. But if v_a in eqn (5.2.2) is replaced by $\hat{v}_a$, then the right-hand side is unchanged since the velocities of the points $P_1, \ldots, P_k$ are unchanged. Therefore $J_a v_a = J_a \hat{v}_a$.

From the impulse equations, $T_{ab} v_b = J_a$. Hence

$$T_{ab} v_a v_b = T_{ab} \hat{v}_a v_b \tag{5.2.3}$$

and therefore

$$\begin{aligned} \tfrac{1}{2}T_{ab}(\hat{v}_a\hat{v}_b - v_a v_b) &= \tfrac{1}{2}T_{ab}(\hat{v}_a\hat{v}_b - 2v_a\hat{v}_b + v_a v_b) \\ &= \tfrac{1}{2}T_{ab}(v_a - \hat{v}_a)(v_b - \hat{v}_b) \\ &\geqslant 0 \end{aligned} \tag{5.2.4}$$

since T_{ab} is positive definite.

Note that there is equality only in the case $v_a = \hat{v}_a$.

Example (5.2.1).† Four equal uniform rods of mass m are smoothly jointed together to form a square $ABCD$ lying at rest on a smooth horizontal table. An impulse is applied at A. Immediately afterwards, A has velocity V in the direction AC (Fig. 5.2.1). The problem is to find the kinetic energy generated by the impulse; and to show that it is less, by $mV^2/30$, than the kinetic energy which the system acquires when the point C is smoothly fixed to the table and the system is set in motion by an impulse at A which, as before, gives A velocity V.

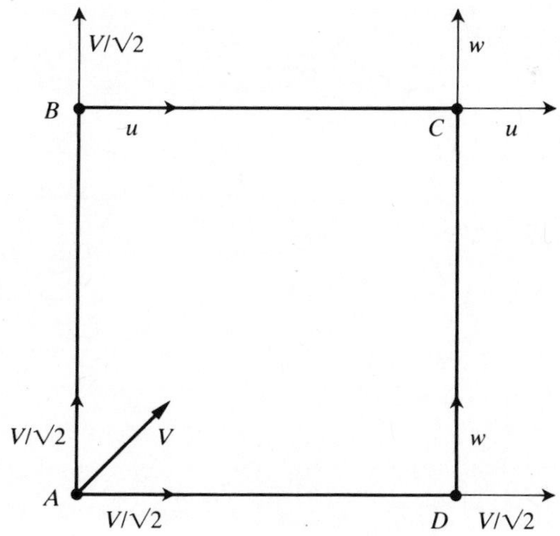

Fig. 5.2.1

Let the velocities of A, B, C, and D immediately after the impulse be as in Fig. 5.2.1. Then, after the impulse,

$$T = \frac{m}{6}(5V^2 + 5u^2 + 5w^2 + \sqrt{2}V(u + w)). \tag{5.2.5}$$

By Kelvin's theorem, this must be minimal with respect to u and w. Hence $\partial T/\partial u = \partial T/\partial w = 0$, which gives

$$u = w = -V/5\sqrt{2}. \tag{5.2.6}$$

The initial kinetic energy is therefore $T = \frac{4}{5}mV^2$.

With $u = w = 0$, $T = \frac{5}{6}mV^2$, which is more by $\frac{1}{30}mV^2$.

Exercise

(5.2.1)† A thin uniform disc has mass M and centre O. A thin uniform rod PQ of mass m is freely hinged at a point P on the circumference. The system lies at rest on a smooth horizontal table with OPQ collinear. An impulse is applied at Q which gives it a velocity v in a horizontal direction perpendicular to PQ. Show that the kinetic energy generated is

$$\frac{m(4M + 3m)v^2}{24(M + m)}.$$

6 Oscillations

6.1 Normal coordinates

In this chapter, we shall look at an analogue of simple harmonic motion in systems with many degrees of freedom. This is not only of interest in itself: it also gives a good approximation to the behaviour near equilibrium of a general class of mechanical systems.

With just one degree of freedom, one can characterize simple harmonic motion by the form of the Lagrangian

$$L = \tfrac{1}{2}v^2 - \tfrac{1}{2}\omega^2 q^2, \tag{6.1.1}$$

which generates the equation of motion

$$\ddot{q} + \omega^2 q = 0. \tag{6.1.2}$$

The constant ω is the *angular frequency* of the oscillations. Angular frequency is measured in radians per second; it is related to the *frequency* ν, which is measured in hertz, or cycles per second, by $\omega = 2\pi\nu$.

In a system with n degrees of freedom the analogue is the motion generated by

$$L = \tfrac{1}{2}K_{ab}v_a v_b - \tfrac{1}{2}P_{ab}q_a q_b, \tag{6.1.3}$$

where the P_{ab} are the entries in a constant $n \times n$ symmetric matrix P and the K_{ab} are the entries in a constant positive definite symmetric matrix K. The corresponding equations of motion are

$$K_{ab}\ddot{q}_b + P_{ab}q_b = 0 \tag{6.1.4}$$

or, equivalently,

$$K\ddot{q} + Pq = 0 \tag{6.1.4'}$$

where q is the column vector with entries q_a.

The key property of eqn (6.1.4) is *linearity*: if $q_a = x_a(t)$, $q_a = y_a(t), \ldots$ are solutions, then so is the linear combination

$$q_a = \alpha x_a(t) + \beta y_a(t) + \cdots \tag{6.1.5}$$

for any constant $\alpha, \beta, \ldots$. By exploiting linearity, we shall break up the general solution, which can be a complicated orbit in configuration space, into a superposition of *fundamental solutions* which have a more straightforward time-dependence. First, however, we need a result from linear algebra.

Proposition (6.1.1). Let K and P be real symmetric $n \times n$ matrices and let K be positive definite. Then there exists a nonsingular matrix B such that $B^t KB$ is the identity matrix and such that $D = B^t VB$ is a diagonal matrix.

Proof. Since K is symmetric, there exists an orthogonal matrix L such that

$$L^t KL = \begin{pmatrix} k_1 & 0 & 0 \cdots 0 \\ 0 & k_2 & 0 \cdots 0 \\ \vdots & & \vdots \\ 0 & 0 & 0 \cdots k_n \end{pmatrix}. \tag{6.1.6}$$

Moreover, since K is positive definite, the k_a are all positive.

Let M be the diagonal matrix with diagonal entries $k_1^{-1/2}, \ldots, k_n^{-1/2}$. Then $M^t L^t KLM = I$ (the $n \times n$ identity matrix); and $S = M^t L^t PLM$ is a symmetric matrix.

Let N be an orthogonal matrix such that $N^t SN$ is diagonal. Put $D = N^t SN$ and $B = LMN$. Then B is non-singular and

$$\begin{aligned} B^t KB &= N^t M^t L^t KLMN = N^t N = I \\ B^t PB &= N^t SN = D \end{aligned} \tag{6.1.7}$$

which completes the proof. □

In index notation, eqns (6.1.7) read

$$B_{ca} K_{cd} B_{db} = \delta_{ab} \quad \text{and} \quad B_{ca} P_{cd} B_{db} = D_{ab} \tag{6.1.8}$$

where $D_{ab} = 0$ when $a \neq b$. If we make a linear transformation to a new system of coordinates $\tilde{q}_a$ which are related to the q_a's by

$$q_a = B_{ab} \tilde{q}_b, \tag{6.1.9}$$

then the Lagrangian becomes

$$\begin{aligned} L &= \tfrac{1}{2} \tilde{v}_a \tilde{v}_a - \tfrac{1}{2} D_{ab} \tilde{q}_a \tilde{q}_b \\ &= \tfrac{1}{2}(\tilde{v}_1^2 + \tilde{v}_2 + \cdots + \tilde{v}_n^2) - \tfrac{1}{2}(\lambda_1 \tilde{q}_1^2 + \lambda_2 \tilde{q}_2^2 + \cdots + \lambda_n \tilde{q}_n^2) \end{aligned} \tag{6.1.10}$$

where $\lambda_1 = D_{11}$, $\lambda_2 = D_{22}$, and so on.

Definition (6.1.1). Generalized coordinates in which the Lagrangian takes the form (6.1.10) are called *normal coordinates*.

In normal coordinates, the equations of motion are

$$\ddot{\tilde{q}}_1 + \lambda_1 \tilde{q}_1 = 0, \qquad \ddot{\tilde{q}}_2 + \lambda_2 \tilde{q}_2 = 0, \ldots \tag{6.1.11}$$

and the dynamical problem is very much simpler: the system of simultaneous linear differential equations for the q_a has been transformed into n separate equations for the $\tilde{q}_a$, which can be solved independently of each other.

The general solution of

$$\ddot{q} + \lambda q = 0 \tag{6.1.12}$$

takes a different form according to whether λ is positive, negative, or zero:

$$\begin{aligned} q &= E \cos \omega t + F \sin \omega t & (\lambda = \omega^2 > 0) \\ q &= E\mathrm{e}^{\omega t} + F\mathrm{e}^{-\omega t} & (\lambda = -\omega^2 < 0) \\ q &= E + Ft & (\lambda = 0) \end{aligned} \tag{6.1.13}$$

(where E and F are constants). Thus in the general solution of the eqns (6.1.11), the time-dependence of each normal coordinate is of one of these forms, depending on the sign of the corresponding λ_a. There are, in particular, special solutions in which all but one of the normal coordinates vanish identically; and every solution is a linear combination of these special solutions.

Let us consider how one of the special solutions looks in the original coordinates q_a. Suppose, for example, $\tilde{q}_2 = \cdots = \tilde{q}_n = 0$ throughout the motion and that $\tilde{q}_1$ is of one of the three forms (6.1.13), with $\lambda = \lambda_1$. Then

$$\begin{aligned} q_a &= (E \cos \omega t + F \sin \omega t) A_a & (\text{if } \lambda = \lambda_1 = \omega^2 > 0) \\ q_a &= (E\mathrm{e}^{\omega t} + F\mathrm{e}^{-\omega t}) A_a & (\text{if } \lambda = \lambda_1 = -\omega^2 < 0) \\ q_a &= (E + Ft) A_a & (\text{if } \lambda = \lambda_1 = 0) \end{aligned} \tag{6.1.14}$$

where the A_a's are the entries in the first column of the matrix B. On substituting each of these into the equation of motion (6.1.4), we find that

$$(\lambda K_{ab} - P_{ab}) A_b = 0. \tag{6.1.15}$$

In all three cases, therefore, there exists a column vector A (which cannot be zero since B is nonsingular), such that

$$(\lambda K - P) A = 0. \tag{6.1.16}$$

It follows that the matrix $\lambda K - P$ is singular and that λ must be one of the roots of the *characteristic equation*

$$\det(\lambda K - P) = 0 \tag{6.1.17}$$

which is a polynomial equation of degree n in λ.

Proposition (6.1.2). The roots of the characteristic equation are the diagonal entries in D. In particular, the roots are all real.

Proof. In the notation of proposition (6.1.1),

$$\begin{aligned}(\lambda-\lambda_1)(\lambda-\lambda_2)\cdots(\lambda-\lambda_n) &= \det(\lambda I - D)\\ &= \det(B^{t}(\lambda K - P)B) \qquad (6.1.18)\\ &= \det(\lambda K - P)\det(B)^2.\end{aligned}$$

The proposition follows since $\det(B) \neq 0$. □

The point of all this is that it gives us a way of finding solutions in the form of eqn (6.1.14) *without actually carrying out the transformation to normal coordinates.* All that we need to do is the following. First solve the characteristic equation. Next, for each root λ, choose two constants E and F and a nonzero column vector A (with entries A_a) such that eqn (6.1.16) holds. Finally, define q_a as a function of t by the appropriate expression in (6.1.14). Then $q_a = q_a(t)$ is a solution. Moreover the existence of the transformation to normal coordinates tells us that the general solution is a linear combination of solutions of this form.

Definition (6.1.2). A solution of one of the forms

$$\begin{aligned} q_a &= (E\cos\omega t + F\sin\omega t)A_a\\ q_a &= (Ee^{\omega t} + Fe^{-\omega t})A_a \qquad (6.1.19)\\ q_a &= (E + Ft)A_a \end{aligned}$$

(where E, F, ω, and A_a are constant) is called a *fundamental solution.* The three types are said to be, respectively, *oscillatory, exponential,* and *linear.*

Oscillatory fundamental solutions are also called *normal modes of oscillation*; and the corresponding angular frequencies $\omega = \sqrt{\lambda}$ are called the *normal frequencies.*

Note that a linear combination of two fundamental solutions corresponding to roots λ and μ can itself be a fundamental solution only if $\lambda = \mu$.

If the characteristic equation has any non-positive roots, then, except for very special choices of initial conditions, the solution contains terms that grow exponentially or linearly with time and the system is *unstable*; but if all the roots are positive, then the system is *stable* and the general motion is a superposition of simple harmonic motions at the normal frequencies.

Example (6.1.1) Lissajous figures. Take $n=2$, $\lambda_1=M^2$ and $\lambda_2=N^2$. Then the general solution in normal coordinates is

$$\tilde{q}_1 = C_1 \cos(Mt+\varepsilon_1), \qquad \tilde{q}_2 = C_2 \cos(Nt+\varepsilon_2), \tag{6.1.20}$$

where C_1, C_2, ε_1, and ε_2 are constants.

The orbits in C are called *Lissajous figures*. An orbit returns eventually to its initial configuration only if M/N is rational; see Fig. 6.1.1.

When N is an integer, $M=1$, $C_1=C_2=1$, and $\varepsilon_1=\varepsilon_2=0$, we have

$$\tilde{q}_1 = \cos t, \qquad \tilde{q}_2 = \cos Nt. \tag{6.1.21}$$

By using trigonometric identities,

$$\tilde{q}_2 = \cos(N \cos^{-1}\tilde{q}_1) = T_N(\tilde{q}_1) \tag{6.1.22}$$

where T_N is a polynomial of degree N in $\tilde{q}_1$, called a *Chebyshev polynomial*. In this case, the configuration oscillates back and forth along the graph of T_N between $\tilde{q}_1=-1$ and $\tilde{q}_1=1$. If the values of ε_1 and ε_2 are changed slightly, then the Lissajous figure becomes a closed curve near the graph of T_N (Fig. 6.1.2). □

Every solution in which all but one of the normal coordinates vanish identically is a fundamental solution. But the converse is not always true.

Consider a fundamental solution corresponding to a root λ of the characteristic equation. Then the time-dependence of each of the q_a's and of each of the $\tilde{q}_a$'s is of the form of eqn (6.1.13). But the normal coordinates $\tilde{q}_a$ must also satisfy eqns (6.1.11), whatever the motion. This is a contradiction unless $\tilde{q}_a$ vanishes identically for every a for which $\lambda_a \neq \lambda$. If all the roots of the characteristic equation are distinct, then the λ_a's are distinct and the fundamental solutions are precisely the solutions in which all but one of the normal coordinates vanish identically. However, if, for example, $\lambda_1=\lambda_2$, then it is possible to have a fundamental solution in which both $\tilde{q}_1$ and $\tilde{q}_2$ are excited.

Example (6.1.2). Consider the Lagrangian

$$L = \tfrac{1}{2}(v_1^2+v_2^2) - \tfrac{1}{2}(q_1^2+q_2^2). \tag{6.1.23}$$

Here q_1 and q_2 are a system of normal coordinates; but $q_1=\cos t$, $q_2=\cos t$ is a fundamental solution in which both q_1 and q_2 are excited. □

The procedure for finding normal coordinates is the same as that for diagonalizing a symmetric matrix, except that the role of the identity

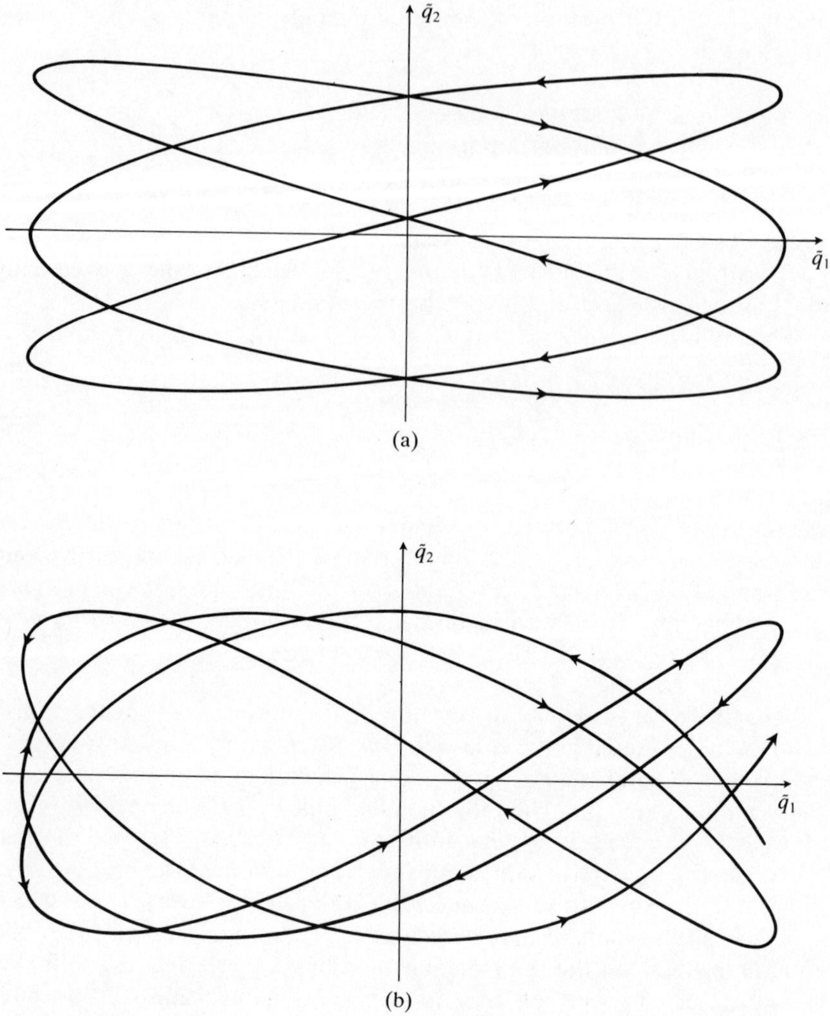

Fig. 6.1.1 Orbits in C: case (a), M/N rational; case (b), M/N irrational.

matrix is filled by the positive definite matrix K. It follows from eqn (6.1.7) that if A is the kth column of B and λ is the kth diagonal entry in D, then

$$(\lambda K - P)A = 0. \tag{6.1.24}$$

To find the colums of B, therefore, we must first find the roots of the characteristic equation and choose corresponding solutions of eqns (6.1.24).

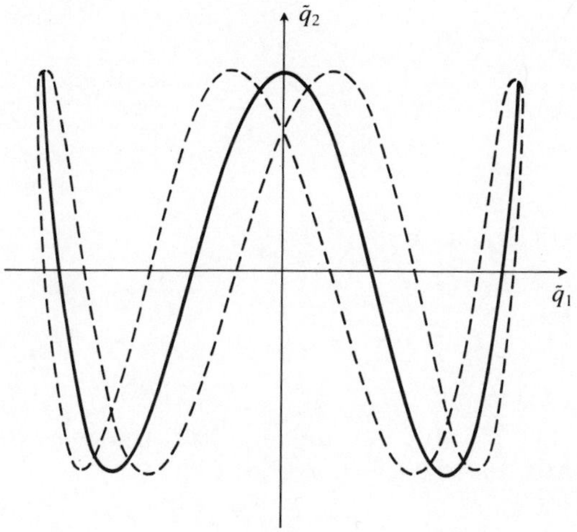

Fig. 6.1.2 The solid curve is the graph T_3. The broken curve is a Lissajous figure.

The choices cannot be made arbitrarily, however: to have $B^tKB = I$, we must choose the different column vectors A so that they are *orthonormal* with respect to K; that is, they are normalized in the sense that

$$A^tKA = 1 \tag{6.1.25}$$

and orthogonal in the sense that if A and A' are two distinct columns vectors (with $\lambda KA = PA$ and $\lambda' KA' = PA'$), then

$$A^tKA' = 0. \tag{6.1.26}$$

If $\lambda \neq \lambda'$, then this latter condition is not a problem: it is an automatic consequence of

$$\begin{aligned}(\lambda - \lambda')A^tKA' &= (\lambda KA)^tA' - A^t(\lambda' KA')\\ &= (PA)^tA' - A^tPA' = 0.\end{aligned} \tag{6.1.27}$$

However, if m of the roots are all equal to λ, say, then the solution space of

$$(\lambda K - P)A = 0 \tag{6.1.28}$$

is m-dimensional; and one must choose the corresponding A's to be an orthonormal basis of the solution space.

Example (6.1.3). Take $n = 3$ and

$$\begin{aligned}T &= \tfrac{1}{2}(2v_1^2 + v_2^2 + 3v_3^2 - 2v_1v_2 - 4v_1v_3 + 2v_2v_3)\\ U &= \tfrac{1}{2}(3q_1^2 + 2q_2^2 + 4q_3^2 - 4q_1q_2 - 6q_1q_3 + 4q_2q_3)\end{aligned} \tag{6.1.29}$$

Then

$$K=\begin{pmatrix}2 & -1 & -2\\ -1 & 1 & 1\\ -2 & 1 & 3\end{pmatrix},\qquad P=\begin{pmatrix}3 & -2 & -3\\ -2 & 2 & 2\\ -3 & 2 & 4\end{pmatrix}. \tag{6.1.30}$$

The characteristic equation is

$$\det(\lambda K-P)=\begin{vmatrix}2\lambda-3 & -\lambda+2 & -2\lambda+3\\ -\lambda+2 & \lambda-2 & \lambda-2\\ -2\lambda+3 & \lambda-2 & 3\lambda-4\end{vmatrix}=0. \tag{6.1.31}$$

Without expanding the determinant, we can spot that $\lambda=1$ is a repeated root and that the third root is $\lambda=2$.

With $\lambda=1$, the solutions of $(\lambda K-P)A=0$ are

$$A=\alpha\begin{pmatrix}1\\0\\1\end{pmatrix}+\beta\begin{pmatrix}1\\1\\0\end{pmatrix} \tag{6.1.32}$$

for any α and β. With $\lambda=2$, the solutions are

$$A=\gamma\begin{pmatrix}0\\1\\0\end{pmatrix} \tag{6.1.33}$$

for any γ.

If A and A' are as in eqn (6.1.32), then

$$A^{\mathrm{t}}KA'=\alpha\alpha'+\beta\beta' \tag{6.1.34}$$

and if A is given by eqn (6.1.33), then

$$A^{\mathrm{t}}KA=\gamma^2. \tag{6.1.35}$$

Hence we can take the three columns of B to be: first, eqn (6.1.32) with $\alpha=1,\ \beta=0$; second, eqn (6.1.32) with $\alpha=0,\ \beta=1$; and third, eqn (6.1.33) with $\gamma=1$. This gives

$$B=\begin{pmatrix}1 & 1 & 0\\ 0 & 1 & 1\\ 1 & 0 & 0\end{pmatrix}. \tag{6.1.36}$$

The corresponding normal coordinates are given by $\tilde{q}=B^{-1}q$. That is

$$\begin{aligned}\tilde{q}_1&=q_3,\\ \tilde{q}_2&=q_1-q_3,\\ \tilde{q}_3&=-q_1+q_2+q_3.\end{aligned} \tag{6.1.37}$$

However, B is not unique: we could take different combinations of α and β for the first two columns.

Exercises

(6.1.1) In example (6.1.3), show that the orbits lie in surfaces in C with equations of the form

$$a\tilde{q}_1^2 + 2b\tilde{q}_1\tilde{q}_2 + c\tilde{q}_2^2 = 1.$$

Sketch some typical orbits on the cylinder $\tilde{q}_1^2 + \tilde{q}_2^2 = 1$.

(6.1.2) Solve the characteristic equation and find normal coordinates in the case that

$$T = \tfrac{1}{2}(3v_1^2 + 3v_2^2 + 3v_3^2 - 2v_1v_2 - 2v_2v_3 - 2v_3v_1)$$
$$U = 3q_1^2 + 3q_2^2 + 3q_3^2 - 2q_1q_2 - 4q_1q_3.$$

6.2 Oscillations near equilibrium

So far there has been no approximation. But now consider a system with Lagrangian

$$L = \tfrac{1}{2}T_{ab}v_av_b - U, \tag{6.2.1}$$

in which U and T_{ab} are functions of the generalized coordinates q_a (but not t). The equilibrium configurations of the system—the configurations in which the Lagrange's equations admit solutions of the form $q_a =$ constant—are given by $\partial U/\partial q_a = 0$.

We are interested in how the system behaves when it is close to equilibrium.

Suppose that the coordinates have been chosen so that $q_a = 0$ is an equilibrium configuration. There is no loss of generality in supposing that $U(0) = 0$. Then, by ignoring terms of the third and higher orders in the coordinates and velocities,

$$L = \tfrac{1}{2}K_{ab}v_av_b - \tfrac{1}{2}P_{ab}q_aq_b \tag{6.2.2}$$

where $K_{ab} = T_{ab}(0)$ and

$$P_{ab} = \left.\frac{\partial^2 U}{\partial q_a\, \partial q_b}\right|_{q=0}. \tag{6.2.3}$$

The motions close to equilibrium are those in which both the q_a's and the v_a's remain small. We should be able to analyse these by replacing the original Lagrangian by its approximation (eqn 6.2.2), although it is far from obvious that the exact solutions of the approximate equations of motion are in fact approximate solutions of the exact equations, particularly over a long period of time. We shall make no attempt to unravel this problem beyond noting that the method is clearly *invalid* if the characteristic equation of (6.2.2) has any non-positive roots: there are then fundamental solutions containing exponential or linear terms, which certainly do not remain close to the equilibrium configuration.

If there are negative roots, then the original system is unstable; and if all the roots are positive, then it is stable. The presence of zero roots, on the other hand, indicates no more than that the approximation is invalid; the stability question is left unresolved.

The following example indicates how the analysis goes in practice.

Example (6.2.1).† A uniform rod of mass m and length $2L$ is suspended by two light elastic strings of natural length a and modulus $\frac{1}{2}mg$, as shown in Fig. 6.2.1. It is free to swing in the vertical plane through the points of suspension. The problem is to find the normal modes of oscillation near equilibrium.

In equilibrium, the strings have length $2a$. Take the equilibrium position of A as the origin and introduce axes with the x-axis horizontal and the y-axis vertically upwards. Let q_1 and q_2 be, respectively, the y and x coordinates of A; and let q_3 and q_4+2L by the y and x coordinates of B.

In fact the system has only three degrees of freedom since q_4 can be eliminated by using the constraint equation

$$(2L+q_4-q_2)^2+(q_3-q_1)^2=4L^2 \tag{6.2.4}$$

which fixes the length of the rod. This reduces to

$$q_4=q_2+O(q^2) \tag{6.2.5}$$

in the first-order approximation.

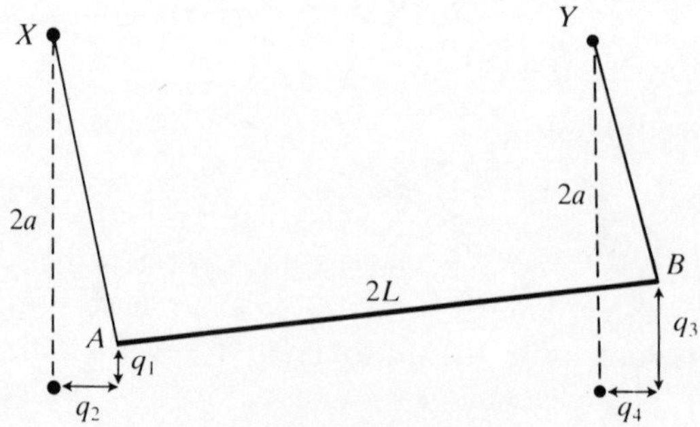

Fig. 6.2.1

The height of the centre of the rod above its equilibrium position is $\frac{1}{2}(q_1+q_3)$; and the lengths of the strings XA and YB are

$$XA=\sqrt{\{(2a-q_1)^2+q_2^2\}}, \qquad YB=\sqrt{\{(2a-q_3)^2+q_4^2\}}. \quad (6.2.6)$$

Hence the potential, which is the sum of the elastic and gravitational potential energies, is

$$U=\tfrac{1}{2}mg(q_1+q_3)+\frac{mg}{4a}((XA-a)^2+(YB-a)^2)-\frac{mga}{2}, \quad (6.2.7)$$

the final constant being chosen to make $U(0)=0$.

The theory demands that we should express U as a function of q_1, q_2, and q_3 and then calculate the entries in the matrix P by taking the second partial derivatives. Of course, this works; but it is not quick. It is simpler to expand U up to the second order in the q_a. From eqn (6.2.6),

$$\begin{aligned} XA &= 2a\sqrt{\left(1-\frac{q_1}{a}+\frac{q_2^2+q_1^2}{4a^2}\right)} \\ &= 2a-q_1+\frac{q_2^2}{4a}+O(q^3) \end{aligned} \quad (6.2.8)$$

with a similar expression for YB. Therefore

$$(XA-a)^2=a^2-2aq_1+\tfrac{1}{2}q_2^2+q_1^2+O(q^3) \quad (6.2.9)$$

Hence, to the second order in q_1, q_2, and q_3,

$$\begin{aligned} U &= \frac{mg}{4a}(q_1^2+\tfrac{1}{2}q_2^2+q_3^2+\tfrac{1}{2}q_4^2) \\ &= \frac{mg}{4a}(q_1^2+q_2^2+q_3^2) \end{aligned} \quad (6.2.10)$$

since $q_4=q_2+O(q^2)$. (We have been lucky: if the expression for U had contained linear terms in q_4, then we would not have got away with the substitution $q_4=q_2$.)

The kinetic energy is easier to deal with: we have only to write down an expression for T *in the equilibrium configuration.* In equilibrium, we have $v_2=v_4$ (exactly) and

$$T=\tfrac{1}{6}m(v_1^2+3v_2^2+v_3^2+v_1v_3) \quad (6.2.11)$$

from the formula for the kinetic energy of a rod (exercise (3.1.4)).

The characteristic equation is

$$\begin{vmatrix} 2a\lambda-3g & 0 & a\lambda \\ 0 & 6a\lambda-3g & 0 \\ a\lambda & 0 & 2a\lambda-3g \end{vmatrix}=0. \quad (6.2.12)$$

Hence the normal (angular) frequencies are $\omega_1 = \sqrt{(g/a)}$, $\omega_2 = \sqrt{(g/2a)}$, and $\omega_3 = \sqrt{(3g/a)}$. The corresponding normal modes are

(1)

$$\begin{pmatrix} q_1 \\ q_2 \\ q_3 \end{pmatrix} = C\cos(\omega_1 t + \varepsilon)\begin{pmatrix} 1 \\ 0 \\ 1 \end{pmatrix} \tag{6.2.13}$$

(C and ε are constants). The strings remain vertical and the two ends of the rod oscillate up and down in phase.

(2)

$$\begin{pmatrix} q_1 \\ q_2 \\ q_3 \end{pmatrix} = C\cos(\omega_2 t + \varepsilon)\begin{pmatrix} 0 \\ 1 \\ 0 \end{pmatrix}. \tag{6.2.14}$$

The lengths of the strings are constant and the rod swings back and forth in the vertical plane.

(3)

$$\begin{pmatrix} q_1 \\ q_2 \\ q_3 \end{pmatrix} = C\cos(\omega_3 t + \varepsilon)\begin{pmatrix} 1 \\ 0 \\ -1 \end{pmatrix}. \tag{6.2.15}$$

The strings remain vertical and the two ends oscillate up and down, exactly out of phase.

Exercises

(6.2.1)† A uniform rod AB of mass m and length $2a$ is suspended by two light inextensible strings PA and QB, each of length $2a$. The fixed points P and Q are at the same level, distance $2a$ apart; and in equilibrium, PA and QB are vertical. The system performs small oscillations about its equilibrium configuration. Determine the normal frequencies and describe the normal modes.

(6.2.2)† Four uniform rods AB, BC, CD, DA, each of mass m and length $2a$, are smoothly jointed together at A, B, C, and D. They are suspended by four light inextensible strings PA, QB, RC, SD, each of length a, attached to fixed points P, Q, R, and S. In equilibrium the strings are vertical and the rods lie in a horizontal plane in the form of a square. Show that the normal frequencies are $\sqrt{(g/a)}$ and $\sqrt{(3g/2a)}$. Describe the normal modes.

(6.2.3)† A particle of mass m is attached by identical light elastic strings of natural length a and modulus of elasticty λ to four points A, B, C, and D, which lie at the corners of a square of side $2a$ in a horizontal plane. In equilibrium, the particle hangs under gravity at a distance $a\sqrt{2}$ below $ABCD$. Show that $mg = \lambda\sqrt{8}$.

Find the normal frequencies for small oscillations about equilibrium and describe the normal modes.

Notes on some of the exercises

(1.1.1) Solutions: (a) 1, (b) 3, (c) 2, (d) 4.

(1.4.3) Put $\boldsymbol{r} = \alpha\boldsymbol{\omega} + \beta\hat{D}\boldsymbol{\omega} + \gamma\boldsymbol{\omega} \wedge \hat{D}\boldsymbol{\omega}$. Show that if $\hat{D}\boldsymbol{\omega} \wedge \boldsymbol{r} + \boldsymbol{\omega} \wedge (\boldsymbol{\omega} \wedge \boldsymbol{r}) = 0$, then $\alpha = \beta = \gamma = 0$.

(1.4.5) Consider the motion relative to the frame $(O, (\boldsymbol{i}, \boldsymbol{j}, \boldsymbol{k}))$. The radius is $ab\Omega(a^2(n - \Omega)^2 + b^2\Omega^2)^{-1/2}$.

(2.3.5) For the first part, multiply the equation of motion by v_a and sum over a.

(2.4.4) Let O be the midpoint of AB. Take the x-axis along OB and the y-axis vertically upwards. Then the coordinates of P are $x = b \cosh \varphi \cos \theta$, $y = -\sinh \varphi \sin \theta$.

(2.5.4) For the last part, use eqn (2.5.19); alternatively, introduce spherical polar coordinates and use the result of exercise (2.5.1).

(2.5.5) For the second part, differentiate eqn (2.5.13) with respect to v_b.

(2.6.3) Write eqn (2.6.9) in the form

$$\frac{\partial^2 F}{\partial v_a\, \partial q_b} v_b + \frac{\partial^2 F}{\partial v_a\, \partial v_b} \dot{v}_b + \frac{\partial^2 F}{\partial v_a\, \partial t} - \frac{\partial F}{\partial q_a} = 0. \qquad (*)$$

This holds for all values of q_a, v_a, $\dot{v}_a$, and t. Deduce that $\partial^2 F/\partial v_a\, \partial v_b = 0$ and hence that $F = A_a(q, t)v_a + B(q, t)$. Substitute into $(*)$ to get

$$\frac{\partial A_a}{\partial q_b} = \frac{\partial A_b}{\partial q_a}, \qquad \frac{\partial A_a}{\partial t} = \frac{\partial B}{\partial q_a}.$$

These imply that $A_a = \partial f/\partial q_a$ and $B = \partial f/\partial t$ for some function $f = f(q, t)$.

(2.6.4) Use the polar angles θ and φ as coordinates. Without loss of generality, choose the polar axis so that $\theta = \pi/2$, $\dot{\theta} = 0$ initially.

(3.1.4) Write

$$T = \int_0^1 \tfrac{1}{2} m(t\boldsymbol{u} + (1 - t)\boldsymbol{v}) \,.\, (t\boldsymbol{u} + (1 - t)\boldsymbol{v})\, \mathrm{d}t.$$

(3.1.6) For the last part, consider a circular disc of radius a and find a point of inertial symmetry on the axis of the disc.

(3.2.1) Substitute $\omega_2 = K \tanh u$, where K is an appropriate constant.

(3.2.2) Use eqns (3.2.13) and (3.2.14) to write ω_1 and ω_3 in terms of T, M^2, and ω_2. Now substitute for ω_1 and ω_3 in $B^2\dot{\omega}_2^2 = (A-C)^2\omega_3^2\omega_1^2$.

(3.2.3) The intersection of the instantaneous axis with the surface has coordinates $(\lambda\omega_1, \lambda\omega_2, \lambda\omega_3)$ where $\lambda = \pm k/\sqrt{(2T)}$. The tangent planes at these points are given by

$$A\omega_1 x + B\omega_2 y + C\omega_3 z = \lambda(A\omega_1^2 + B\omega_2^2 + C\omega_3^2)$$

or, alternatively, by $\boldsymbol{M}_O \cdot \boldsymbol{r} = \pm\sqrt{(2T)}k$. But $\boldsymbol{M}_O$ is fixed relative to the inertial frame and T is constant.

The shape of the surface is not important: the ellipsoid can be an imaginary surface in the body with equation $Ax^2 + By^2 + Cz^2 = k^2$. The motion is then such that the imaginary surface appears to be rolling between two fixed plane—a result due to Poinsot.

(3.2.5) The angular momentum of the smaller sphere about its centre of mass is $\frac{2}{5}ma^2\boldsymbol{\omega}$ where m is its mass and $\boldsymbol{\omega}$ is its angular velocity. Hence if $\boldsymbol{R}$ is the force at the point of contact, then

$$ma\ddot{\boldsymbol{e}} = -mg\boldsymbol{k} + \boldsymbol{R}, \qquad \tfrac{2}{5}ma^2\dot{\boldsymbol{\omega}} = a\boldsymbol{e} \wedge \boldsymbol{R}$$

by the principles of linear and angular momentum. The rolling condition at the point of contact is $a\dot{\boldsymbol{e}} + \boldsymbol{\omega} \wedge (a\boldsymbol{e}) = 0$.

(3.2.6) C. E. Easthope[8] gives a full discussion in his *Three dimensional dynamics*; he also makes an interesting remark about golf.

Let the mass of the sphere be m; let its radius be a and let the radius of the cylinder be $a + c$. The centre of the sphere has position vector $\boldsymbol{r} = z\boldsymbol{k} + c\boldsymbol{e}$ from a fixed origin on the axis of the cylinder, where $\boldsymbol{k}$ and $\boldsymbol{e}$ are orthogonal unit vectors, with $\boldsymbol{k}$ pointing vertically upwards. Let $\boldsymbol{f} = \boldsymbol{k} \wedge \boldsymbol{e}$.

The triad $(\boldsymbol{e}, \boldsymbol{f}, \boldsymbol{k})$ is orthonormal and has angular velocity $\Omega\boldsymbol{k}$ with respect to fixed axes, where Ω is some function of time.

Let $\boldsymbol{\omega}$ be the angular velocity of the sphere. Put $n = \boldsymbol{\omega} \cdot \boldsymbol{e}$ and $N = \boldsymbol{\omega} \cdot \boldsymbol{f}$.

With the dot denoting the time derivative with respect to fixed axes, the equations of motion are

$$\tfrac{2}{5}ma^2\dot{\boldsymbol{\omega}} = a\boldsymbol{e} \wedge \boldsymbol{R}, \qquad m(\ddot{z}\boldsymbol{k} + c\ddot{\boldsymbol{e}}) = \boldsymbol{R} - mg\boldsymbol{k}$$

where $\boldsymbol{R}$ is the force at the point of contact. The rolling condition is

$$\dot{z}\boldsymbol{k} + c\dot{\boldsymbol{e}} + a\boldsymbol{\omega} \wedge \boldsymbol{e} = 0.$$

The following steps lead to the stated result.

(1) From all three equations

$$\tfrac{2}{5}a\dot{\boldsymbol{\omega}} = \boldsymbol{e} \wedge (-a\dot{\boldsymbol{\omega}} \wedge \boldsymbol{e} - a\boldsymbol{\omega} \wedge \dot{\boldsymbol{e}} + g\boldsymbol{k}).$$

Hence $\boldsymbol{e} \cdot \dot{\boldsymbol{\omega}} = 0$ and $\frac{7}{5}a\dot{\boldsymbol{\omega}} = (\Omega na - g)\boldsymbol{f}$.

(2) From the rolling condition

$$\dot{z} = aN \quad \text{and} \quad 0 = c\Omega + a\boldsymbol{\omega} \cdot \boldsymbol{k}.$$

(3) $\dot{\boldsymbol{\omega}} \cdot \boldsymbol{k} = 0$ and hence Ω is constant.

(4) By considering $\dot{\boldsymbol{\omega}} \cdot \boldsymbol{f}$,

$$7\ddot{N} + 2\Omega\dot{n} = 0.$$

(5) By considering $\dot{\boldsymbol{\omega}} . \boldsymbol{e}$, $\dot{n} = \Omega N$.
(6) $7\ddot{z} + 2\Omega^2 \dot{z} = 0$.

(3.3.1) Part (b): remember that $A\Omega^2 u_0 - Cn\Omega + mga = 0$ and that

$$\Omega = \frac{h - Cnu_0}{A(1 - u_0^2)}.$$

(3.3.3) Use the fact that φ is cyclic to show that

$$\dot{\varphi} = \frac{V \sin \alpha}{a \sin^2 \theta}.$$

Show that

$$\dot{\theta}^2(1 + 3\cos^2\theta) + 4\dot{\varphi}^2 \sin^2\theta$$

is constant. Write $u = \cos\theta$ and find an expression for $\mathrm{d}t/\mathrm{d}u$.

(3.3.4) Let α be the angle between the downward vertical and a line joining the centre of the smaller cylinder to a point on its rim and let x be the horizontal distance of the centre of the larger cylinder from a fixed point. Let $(\boldsymbol{i}, \boldsymbol{j}, \boldsymbol{k})$ be an orthonormal triad, with $\boldsymbol{k}$ vertical and $\boldsymbol{i}$ orthogonal to the axes of the cylinders. Then their angular velocities are $\dot{\theta}\boldsymbol{j}$ (large) and $\dot{\alpha}\boldsymbol{j}$ (small); the velocities of the centres are $\dot{x}\boldsymbol{i}$ (large) and

$$\dot{x}\boldsymbol{i} - (\tfrac{1}{2}a \cos\varphi)\dot{\varphi}\boldsymbol{i} + (\tfrac{1}{2}a \sin\varphi)\dot{\varphi}\boldsymbol{k} \quad \text{(small)}.$$

The vector $\boldsymbol{e} = -\cos\varphi\, \boldsymbol{k} - \sin\varphi\, \boldsymbol{i}$ is a unit vector pointing from the centre of the larger cylinder to the centre of the smaller cylinder. The rolling conditions are

$$\dot{x} - a\dot{\theta} = 0;$$

and

$$\dot{x}\boldsymbol{i} + \dot{\theta}\boldsymbol{j} \wedge (a\boldsymbol{e}) = \dot{x}\boldsymbol{i} - (\tfrac{1}{2}a\cos\varphi)\dot{\varphi}\boldsymbol{i} + (\tfrac{1}{2}a \sin\varphi)\dot{\varphi}\boldsymbol{k} + \dot{\alpha}\boldsymbol{j} \wedge (\tfrac{1}{2}a\boldsymbol{e}),$$

which gives $\dot{\alpha} = 2\dot{\theta} - \dot{\varphi}$ and hence $\alpha = 2\theta - \varphi + \text{constant}$.

Take θ and φ as generalized coordinates, and choose origins such that $x = a\theta$ and $\alpha = 2\theta - \varphi$. Then the total kinetic energy is

$$\tfrac{1}{2}ma^2\dot{\theta}^2 + \tfrac{1}{2}m(a\dot{\theta} - \tfrac{1}{2}a\dot{\varphi}\cos\varphi)^2 + \tfrac{1}{8}ma^2\dot{\varphi}^2\sin^2\varphi + \tfrac{1}{2}ma^2\dot{\theta}^2 + \tfrac{1}{8}ma^2(2\dot{\theta} - \dot{\varphi})^2.$$

The result follows from conservation of energy ($\partial L/\partial t = 0$).

(3.4.3) In the notation of (3.2.6) above: use as coordinates θ, φ, ψ, χ, and z, where θ, φ, and ψ are the Euler angles of a triad fixed in the sphere relative to the triad $(\boldsymbol{e}, \boldsymbol{f}, \boldsymbol{k})$; χ is the angle between $\boldsymbol{e}$ and a fixed horizontal line; and z is the height of the centre of the sphere above a fixed origin on the axis of the cylinder.

Then

$$\boldsymbol{\omega} . \boldsymbol{k} = \dot{\chi} + \dot{\varphi} + \dot{\psi}\cos\theta$$
$$n = -\dot{\theta}\sin\varphi + \dot{\psi}\sin\theta\cos\varphi$$
$$N = \dot{\theta}\cos\varphi + \dot{\psi}\sin\theta\sin\varphi.$$

The Lagrangian is

$$L = \tfrac{1}{5}ma^2(\dot\theta^2 + \dot\psi^2 + (\dot\varphi + \dot\chi)^2 + 2\dot\psi(\dot\varphi + \dot\chi)\cos\theta) + \tfrac{1}{2}m(\dot z^2 + c^2\dot\chi^2) - mgz$$

and the rolling conditions are

$$\dot z - a(\dot\theta\cos\varphi + \dot\psi\sin\theta\sin\varphi) = 0$$
$$c\dot\chi + a(\dot\varphi + \dot\chi + \dot\psi\cos\theta) = 0,$$

corresponding to which we have the Lagrange multipliers μ and λ.

The φ and χ equations give that $\lambda = 0$ and that $\dot\chi = \Omega$ is constant. After some manipulation, the ψ and θ equations give

$$\tfrac{2}{5}ma\ddot\psi\sin\theta + \tfrac{4}{5}ma\dot\psi\dot\theta\cos\theta + \tfrac{2}{5}mc\dot\chi\dot\theta = -\mu\sin\varphi$$
$$\tfrac{2}{5}ma\ddot\theta - \tfrac{2}{5}ma\dot\psi^2\sin\theta\cos\theta - \tfrac{2}{5}mc\dot\chi\dot\psi\sin\theta = -\mu\cos\varphi$$

The z equation is $m\ddot z + mg = \mu$. A little further work leads to $\dot n = \Omega N$ and $7\dot N + 2\Omega n + 5g/a = 0$ and hence $7\dddot z + 2\Omega^2\dot z = 0$, as before.

(4.2.2) Consider

$$\frac{\partial(q, p, t)}{\partial(q', p', t')} \quad \text{and} \quad \frac{\partial(q'', p'', t'')}{\partial(q', p', t')}.$$

(4.3.1) The vector field $\boldsymbol{X}$ is tangent to Σ if and only if

$$\boldsymbol{X}\,.\,\mathrm{grad}\left(p - \frac{\partial S}{\partial q}\right) = 0$$

on Σ, where grad is the gradient operator

$$\mathrm{grad} = \boldsymbol{i}\frac{\partial}{\partial q} + \boldsymbol{j}\frac{\partial}{\partial p} + \boldsymbol{k}\frac{\partial}{\partial t}.$$

This is equivalent to

$$0 = -\frac{\partial h}{\partial q} - \frac{\partial^2 S}{\partial q^2}\frac{\partial h}{\partial p} - \frac{\partial^2 S}{\partial q\,\partial t}$$
$$= -\frac{\partial}{\partial q}\left(\frac{\partial S}{\partial t} + h\left(q, \frac{\partial S}{\partial q}, t\right)\right).$$

(4.3.3) See exercise (2.4.4).

(5.1.5) The last part can be solved by adapting the initial motion technique. Let θ be the angle between AB and a fixed direction in the plane. It is clear that θ is cyclic, so $T = T(\psi, \dot\theta, \dot\psi)$.

There are no external forces and $\dot\psi = 0$ initially (that is immediately after the application of the impulse). Therefore the initial values of $\ddot\theta$ and $\ddot\psi$ are given by

$$0 = \frac{\mathrm{d}}{\mathrm{d}t}\left(\frac{\partial T}{\partial\dot\theta}\right) = \frac{\partial^2 T}{\partial\dot\theta^2}\ddot\theta + \frac{\partial^2 T}{\partial\dot\theta\,\partial\dot\psi}\ddot\psi$$
$$\frac{\partial T}{\partial\psi} = \frac{\mathrm{d}}{\mathrm{d}t}\left(\frac{\partial T}{\partial\dot\psi}\right) = \frac{\partial^2 T}{\partial\dot\psi\,\partial\dot\theta}\ddot\theta + \frac{\partial^2 T}{\partial\dot\psi^2}\ddot\psi.$$

The partial derivatives with respect to $\dot{\theta}$ and $\dot{\psi}$ on the right are taken in the initial configuration with the values of θ and ψ held fixed. They are given by differentiating

$$T = \tfrac{10}{3}ma^2(\dot{\theta}^2 + (\dot{\theta} + \dot{\psi})^2),$$

which is the expression for T when $\psi = \pi/2$.

The derivative $\partial T/\partial\psi$ is taken with the values of $\dot{\theta}$ and $\dot{\psi}$ held fixed at $\dot{\theta} = \omega$ and $\dot{\psi} = 0$. Now when $\dot{\psi} = 0$ the total kinetic energy of each rod is the rotational kinetic energy associated with its angular velocity $\dot{\theta}$, which does not change if we change ψ keeping $\dot{\theta}$ and $\dot{\psi}$ fixed; and the translational kinetic energy of its centre of mass, which is $\frac{1}{2}mL^2\dot{\theta}^2$ where L is the distance of its centre from A. For the rods AD and AB, L is independent of ψ. For the rods BC and DC

$$L^2 = 4a^2\sin^2\psi + (a + 2a\cos\psi)^2.$$

Hence when ψ takes its initial value of $\pi/2$,

$$\frac{\partial T}{\partial\psi} = m\omega^2\frac{\partial}{\partial\psi}(L^2) = -4ma^2\omega^2.$$

(5.1.7) Find the θ, φ, and ψ components of J. Use the impulse equations to find p_θ, p_φ, and p_ψ immediately after the impulse and then use the equations of motion to determine the sign of $\ddot{\theta}$.

(6.1.2) The roots are $\lambda = 1$, $\lambda = 2$, and $\lambda = 3$.

(6.2.1) One might be tempted to begin by introducing four coordinates; the spherical polar angles θ_1 and φ_1 of A (with polar axis PA): and the two spherical polar angles θ_2 and φ_2 of B (with polar axis QB). This will not work, however; the polar coordinates are singular in the equilibrium configuration, as is reflected by the fact that large changes in φ_1 and φ_2 can correspond to small displacements in the system.

(6.2.2) Note that the system has four degrees of freedom.

(6.2.3) Take the origin to be the equilibrium position of the particle, with axes chosen so that the position vectors of A, B, C, and D are $\boldsymbol{a} = a(1, 1, \sqrt{2})$, $\boldsymbol{b} = a(1, -1, \sqrt{2})$, $\boldsymbol{c} = a(-1, -1, \sqrt{2})$, and $\boldsymbol{d} = a(-1, 1, \sqrt{2})$. Suppose that the particle is at the point P with position vector $\boldsymbol{r} = (x, y, z)$. To the second order in x, y, z:

(1)

$$PA = 2a\left(1 - \frac{\boldsymbol{a}.\boldsymbol{r}}{4a^2} + \frac{\boldsymbol{r}.\boldsymbol{r}}{8a^2} - \frac{(\boldsymbol{a}.\boldsymbol{r})^2}{32a^4}\right)$$

(2) the extension of the string PA is

$$\left(a^2 - \boldsymbol{a}.\boldsymbol{r} + \tfrac{1}{2}\boldsymbol{r}.\boldsymbol{r} + \frac{1}{8a^2}(\boldsymbol{a}.\boldsymbol{r})^2\right)^{1/2}$$

(3)

$$(\boldsymbol{a} + \boldsymbol{b} + \boldsymbol{c} + \boldsymbol{d}).\boldsymbol{r} = 4a\sqrt{2}z$$

(4)

$$(\boldsymbol{a}\,.\,\boldsymbol{r})^2+(\boldsymbol{b}\,.\,\boldsymbol{r})^2+(\boldsymbol{c}\,.\,\boldsymbol{r})^2+(\boldsymbol{d}\,.\,\boldsymbol{r})^2=4a^2(x^2+y^2+2z^2).$$

Hence

$$U=\frac{\lambda a}{2}\left(4-\frac{4\sqrt{2}z}{a}+\frac{5x^2}{2a^2}+\frac{5y^2}{2a^2}+\frac{3z^2}{a^2}\right)+mgz.$$

Reference notes

1. E. P. Wigner, The unreasonable effectiveness of mathematics. *Commun. Pure Appl. Math.* **13,** 1–14, 1960.
2. V. I. Arnol'd, *Mathematical methods of classical mechanics* (translated by K. Vogtmann and A. Weinstein). Springer-Verlag, New York, 1978.
3. The quotation is from S. Drake's translation: *Galileo Galilei*: *Dialogue concerning the two chief world systems—Ptolemaic and Copernican.* University of California, Berkeley, 1953.
4. The quotations are from Motte's translation of the *Principia,* updated by F. Cajori: *Isaac Newton*: *The mathematical principles of natural philosophy and his system of the world.* University of California, Berkeley, 1934; they are reproduced in *Newton's philosophy of nature* (ed. H. S. Thayer). Hafner, New York, 1953.
5. J. L. Synge and B. A. Griffith, *Principles of mechanics* (3rd edn: International student edition), McGraw-Hill Kogakusha, Tokyo, 1970.
6. D. W. Jordan and P. Smith, *Nonlinear ordinary differential equations* (Second edition). Oxford University Press, 1987.
7. E. T. Whittaker, *A treatise on the analytical dynamics of particles and rigid bodies.* Cambridge, 1904, with many subsequent editions.
8. C. E. Easthope, *Three dimensional dynamics, a vectorial treatment* (2nd edn). Butterworths, London, 1964.
9. Reprinted by kind permission of the Oxford Delagacy of Local Examinations.

Index

Note: some symbols that retain a fixed meaning for a substantial part of the text have been included.